8/88

D1127485

The Geological Disposal
of Nuclear Waste

The Geological Disposal of Nuclear Waste

Neil A. Chapman

Fluid Processes Research Group, British Geological Survey

and

Ian G. McKinley

Eidg. Institut für Reaktorforschung, Würenlingen, Switzerland

with contributions on radiological protection by

Marion D. Hill

National Radiological Protection Board, United Kingdom

JOHN WILEY & SONS

Chichester · New York · Brisbane · Toronto · Singapore

Copyright © 1987 by John Wiley & Sons Ltd

Library of Congress Cataloging-in-Publication Data:
Chapman, Neil A.
  The geological disposal of nuclear waste.

  Includes index.
  1. Radioactive waste disposal in the ground—Great Britain.   2. Engineering geology—Great Britain.
  3. Geochemistry—Great Britain.   4. Hydrogeology.
  I. McKinley, Ian G.   II. Hill, Marion D.   III. Title.
  TD898.2.C44 1987   621.48'38   86-15970

ISBN 0 471 91249 2

British Library Cataloguing in Publication Data:
Chapman, Neil A.
  The geological disposal of nuclear waste.
  1. Radioactive waste disposal
  I. Title   II. McKinley Ian G.
  III. Hill, Marion D.
  621.48'38   TD898

ISBN 0 471 91249 2

Printed and bound in Great Britain

Contents

Acknowledgements

This book went through many versions before it saw the light of day, and in the process both lost and collected authors over the years. What had started in 1979 as a rash idea to write a fairly short book on disposal of high-level waste in the United Kingdom, gradually gave way to something considerably more complicated, wide-ranging and of course onerous. We soon became aware of the problems of trying to write an up-to-date account of a subject which is moving forwards like a train.

Many people have wittingly or unwittingly helped to get this project moving and to keep both it and us on the rails. First, of course, we must mention our families, who have suffered long at our hands as we burned the midnight (and holiday) oil to wrestle with the pen, neglecting them in the process. To Linda, Jill and Anna we say thank you, and promise that it will be a long time before we try another venture like this!

Then we would like to thank all those who have read, commented on, advised or cajoled us into thinking again about what we were doing. Among these, special mention goes to Ghislain de Marsily and Charles McCombie, who struggled with some later drafts. Many other of our colleagues have given us assistance, and we thank all those with whom we cross paths daily for providing a stimulating atmosphere in which to work; particular mention to our colleagues (and ex-colleagues) in the group which knows itself as 'Flupu'.

Finally, thanks go to all those organizations and individuals who have allowed us to quote freely (and to use tables and diagrams) from their work, from both published and unpublished material. Particular thanks to **BGS, NAGRA, SKB** and **UKAEA**.

Preface

Waste disposal has become a key issue in these environmentally conscious times. Mankind has been disposing of wastes in the ground ever since he first tossed a half-gnawed bone down a cleft in the rock. The principle which has applied since the dawn of time has been 'out of sight, out of mind' and, perhaps surprisingly to many, this has been tacitly accepted by society until only a few years ago. It was not until the early 1970s that the more advanced countries began to think in terms of legislating for the safe disposal of all categories of domestic and industrial waste. By that time there already existed a noxious legacy of poisonous wastes which had been poured or thrown into the oceans, wilderness areas or any convenient hole in the ground, with little or no regard for their long-term safety. Even provisions for the routine management of domestic sewage, one of the more important contributions to public health since the industrial revolution, are by no means universally acceptable, even in the developed countries, and are clearly a major concern in the third world.

The modern concept of waste management means that once we dispose of waste, we should not expect to suffer the consequences of its eventual return to our environment. This was rarely considered in disposing of the ever increasing volumes of garbage and industrial residues produced during the last 100 years. Many countries acted too late to effect any simple or economical solution to the problem. Still others have not yet acted. In the USA, one of the first countries to realize the scale of the problem, the damage was already severe, and the present-day costs of cleaning up the mess produced, largely over the last 50 years, are astronomical. Estimates for remedial action on industrial waste-disposal sites alone reach a figure of at least $100 billion.

Somewhere amongst all this toxic garbage, starting about 40 years ago, we begin to see the arrival of small quantities of wastes recognized as radioactive. Although frequently advanced as 'Man's Ultimate Problem', the scale of the issue has to be seen in the light of other long-lived (or indestructible) wastes produced by a society rushing through one technological revolution after another. In these terms the relative quantities of nuclear wastes are trivial. Early disposals suffered much the same off-hand treatment as any other wastes—there were no regulations and, though it may be hard for us to appreciate now, it did not really seem important at the time. Perhaps the only difference between these disposals and other noxious wastes was that the radioactive wastes were generally disposed of on government-controlled lands, and therefore were largely remote from society.

Since the early 1970s, all types of waste have come under much stricter regulatory control, and radioactive wastes more than any other type; so much so that an enormous international research and development effort has been launched during the last 10 years. The results of this impetus allow us now to describe in this book detailed concepts for the long-term management and disposal of such wastes which would have been unthinkable in 1970. For the scientists and technologists among our readers, it is always worth remembering that the science underlying 'radwaste' management has been based firmly on those disciplines which were the pistols aimed at the toxic industrial waste problem; notably hydrogeology, chemistry and materials science. It is perhaps ironic that their application to radwaste has now far outstripped advances in the latter field. In the words of Delcoigne (1985): 'It is high time we applied to the chemical industry techniques similar to those required for ... nuclear waste'.

In writing this book we have tried to adopt a discursive approach in order to make it more readable, particularly for those with a non-scientific background. It is consequently not intended to be a textbook or compendium of scientific data, and we have avoided detailed mathematical treatments. For those wishing to go into a topic in more detail we have provided specific references throughout and a guide to the key sources of information. It has been said that the literature on this subject is now so large as to constitute a waste-disposal problem in itself, but we hope that in adding to this mountain we have at least the excuse of providing a first comprehensive overview of the scientific rationale and methodology of geological disposal of radioactive wastes.

CHAPTER 1

Radioactivity and Radiological Protection

RADIOACTIVE WASTES

Radioactive wastes arise from almost all activities involving the handling of radioactive materials. They are distinguishable from other types of toxic and industrial wastes only by reason of their enhanced radioactivity and, in some cases, their heat output. Since both the radioactivity and heat output of nuclear waste decrease with time, radioactive material eventually becomes inert to all intents and purposes and consequently as unremarkable as many other types of waste. The wastes can be classified depending on the isotopes present, their concentrations, the intensity of their radioactivity, and the time they take to decay to innocuous levels. It might be borne in mind, for comparative purposes, that certain toxic wastes currently disposed of into the environment (for example, those containing arsenic, cadmium and mercury) do not decay and will always remain toxic, while the concept of decreasing toxicity with time is also utilised in the management of other chemical wastes (e.g. unstable or biodegradable species).

RADIOACTIVITY AND ITS EFFECTS

The starting point of this study of radioactive wastes must be a review of the nature of radioactivity itself, and its effects. Matter, as we all know, is composed of atoms. The atomic nucleus itself is composed of positively charged protons and uncharged neutrons. The number of protons is balanced, in the neutral atom, by an identical number of equally charged electrons which surround the nucleus. Such electrons are responsible for the chemical properties of the atom. All atoms with the same number of protons are chemically identical and are

1

called *isotopes* of a particular *element*. Isotopes are distinguished by the total number of protons and neutrons (collectively termed nucleons) in their nucleus. Conventionally, the number of protons in a nucleus (the atomic number) is represented only by the chemical symbol for the element involved, while the sum of protons and neutrons (the 'mass number') is given as a suffix. For example:

(a) atoms with 6 protons correspond to the element carbon, which has the chemical symbol C;

(b) a nucleus with 6 protons and 6 neutrons is called carbon-12 and is represented as ^{12}C;

(c) a nucleus with 6 protons and 7 neutrons would be ^{13}C and one with 7 protons and 6 neutrons would be ^{13}N (N being the symbol for nitrogen— an atom with 7 protons);

(d) atoms specified in terms of chemical symbol and mass number are called *nuclides*;

(e) atoms having the same chemical symbol (atomic number) but different mass numbers are called isotopes. For instance ^{12}C and ^{13}C are isotopes of carbon.

Some elements have very few isotopes, for example, hydrogen has only three, one of which is radioactive (^{3}H, known as tritium). Others have many isotopes. Caesium (Cs) for example has 31, from ^{116}Cs to ^{146}Cs. Only one of these (^{133}Cs) is stable (i.e. non-radioactive).

Many nuclides are inherently unstable and radioactive. These are called *radionuclides*, and will spontaneously transform to other nuclides (of the same or of a different element) by emission of electromagnetic radiation or particles, or by splitting (fission) of the active nucleus. This transformation (radioactive decay) is a random process and hence can be described only by statistical techniques. Given a sufficiently large number of radioactive nuclei, the fraction which will decay within a set time period can be predicted, but whether an individual nucleus will be involved or not cannot be specified. Simplistically, this can be thought of as similar to mortality statistics—even if it is known that 1 per cent of a large population will die in a particular type of accident each year it is not possible to specify which individuals would be involved. Nuclei which are less stable than others will decay more quickly and this can be characterised statistically by a *half-life*. The half-life is the time required for a large number of atoms of a particular radionuclide to decrease by 50 per cent due to radioactive decay. This may correspond to fractions of a second or many millions of years. Our earlier example of Cs has isotopes with half-lives varying from less than 0.2 seconds up to 2 million years. The characteristic of halving of radionuclide content in a set period of time is known as exponential decay and greatly reduces the number of atoms of the element originally present in an apparently short number of half-lives. For a nuclide with a half-life of 1 year, for example, the number of atoms of the original material would be reduced to less than a thousandth of the original number after 10 years, less than a millionth after 20 years, less than a thousand millionth after 30 years, and so on.

In real life, naturally, things are not always so simple and sometimes the nuclide produced by a decay process is itself unstable and again decays. The decay of 'heavy' nuclides (particularly those with high atomic numbers, greater than about 200) often results in chains of many radioactive 'daughters'. Although the mathematics required to describe such chain decay is rather complex, the same basic principle of exponential decay holds (see standard texts such as Friedlander *et al*; 1981, or Choppin and Rydberg; 1980, for a more comprehensive description). The nuclear transformation involved in the decay process is generally classified in terms of the radiation emitted. For the purposes of this discussion, these radiations can be considered as:

(1) α—the emission of an alpha particle comprising a group of 2 protons and 2 neutrons (equivalent to a ^{4}He nucleus) which decreases the mass number of the decaying nucleus by 4 and its atomic number by 2

(2) β$^-$—the emission of an electron (beta particle) from the nucleus resulting from the conversion of a neutron to a proton (i.e. mass number stays constant but atomic number decreases by 1)

(3) β$^+$—the emission of a positron from the nucleus resulting from the conversion of a proton to a neutron (again mass number is constant but atomic number is increased by 1)

(4) n—the emission of a neutron, generally following fission of the nucleus: the atomic number remains constant, but mass number decreases by one

(5) γ—the emission of high energy electromagnetic radiation due to internal stabilisation of the nucleus without change of the number of protons or neutrons

This list is by no means comprehensive (ignoring, for example, emissions resulting from electronic re-arrangements following decay), but covers the processes of greatest relevance in nuclear waste considerations.

An important factor which distinguishes each of these types of radioactive emission is their power to penetrate matter. The α-particles interact very strongly with matter and hence are easily stopped (e.g. by a thin sheet of paper). Although β-particles interact less strongly with matter, they may be effectively stopped by a thin layer of metal. The n and γ radiations are, however, very penetrating and require large thicknesses of shielding materials, e.g. lead or concrete. It should be noted that penetrating power is inversely proportional to degree of interaction with matter (sometimes termed *linear energy transfer*—LET). The latter is a measure of the intensity of ionization caused (i.e. ejection of electrons from the target atoms) which, in turn, reflects the intensity of damage caused by such 'ionising radiation'.

In the course of our discussions on waste disposal we shall be considering mainly α, β and γ radiation as these are the main sources of activity in the wastes. The 'activity' of any radioactive material is measured in terms of the number of individual nuclei which decay or disintegrate each second. The decay rate of 1 nucleus per second is known as a Bequerel. (The old unit of activity is the Curie; equivalent to 3.7×10^{10} Bequerels). Since these units tell us nothing about the type of radiation, its energy, or its degree of interaction with matter, a further unit, the Gray, will be referred to. This unit is a measure of the dose—the energy absorbed by a material (equivalent to damage caused) when radiation strikes it. One Gray is equivalent to the radiation required to cause a kilogram of material to absorb one joule of energy (and is 100 times larger than the old unit—the rad). If the target of the radiation is human tissue this unit of absorbed energy can be translated into a 'radiation dose equivalent' unit called the Sievert by multiplying by factors depending on the biological effects of the type of incident radiation. Both α and neutron radiation are taken to be twenty times more damaging than β or γ radiation in terms of human biology. The Sievert has replaced the old 'Roentgen equivalent man' unit, or rem, and is 100 times larger.

There are three other quantities which are used to describe radiation dose to human beings. The first of these is the 'effective dose equivalent', which is also expressed in Sieverts. This is the sum of the dose equivalents to each part of the human body, weighted by the factors which take account of the different susceptibilities to radiation-induced harm of the various tissues. Effective dose equivalent is particularly useful because it provides a measure of the total risk to an individual.

The two other quantities are the 'collective effective dose equivalent' and the 'collective effective dose equivalent commitment'; both are expressed in man-Sieverts (man Sv). The collective effective dose equivalent is the sum of the individual effective dose equivalents received by a group of people; that is, it is the total dose to a population in a given time. The collective effective dose equivalent commitment is the total dose to generations of people over all time, and thus represents a measure of the total risk associated with a practice involving radiation exposure of people, now and in the future.

Summarizing, radioactivity is measured in Bequerels (Bq), absorbed dose in Grays (Gy), individual human dose equivalent in Sieverts (Sv), and dose equivalent to populations in man Sieverts (man Sv). The biological effects of radiation are the subject of many easily accessible books and articles (e.g. UNSCEAR, 1982; UNEP, 1985). It is not proposed to go into this field here, other than to say that the harmful effects of radiation are well known to occur when absorbed radiation energy kills individual human cells, prevents them from dividing and reproducing normally, or causes mutation of the genetic information carried in them, which can lead to cancers. Depending on the organ or tissue affected, radiation damage can manifest itself in many medical forms. Owing to the short penetration distance of α-particles in particular, major organs may generally only be affected by ingestion or inhalation of such material. The calculation of doses from drinking waters and foods contaminated by leakage from a disposal site is thus a major part of any safety assessment.

Radioactivity (and hence absorbed dose or human dose equivalent) is considered over a vast range, from the limits of measurement to the levels involved in high-level waste management. For convenience of representation, therefore, the SI Unit multipliers are widely used in the literature (Table 1.1) and will be adopted henceforth in this book.

Table 1.1 The SI order-of-magnitude multipliers

Multiplier	Symbol	Meaning
tera-	T	$\times 10^{12}$
giga-	G	$\times 10^{9}$
mega-	M	$\times 10^{6}$
kilo-	k	$\times 10^{3}$
milli-	m	$\times 10^{-3}$
micro-	μ	$\times 10^{-6}$
nano-	n	$\times 10^{-9}$
pico-	p	$\times 10^{-12}$

RADIOACTIVITY IN THE ENVIRONMENT

In order to put discussion of the quantities of waste considered for geological disposal, and possible resultant future releases of radioactivity into context, it is worth considering the other sources of ionizing radiation in the environment, both natural and artificial. Natural materials do not contain only the stable isotopes of the elements. Since both stable and radioactive isotopes of any element generally behave in the same chemical manner, they are to be expected to occur incorporated together in any material. Consequently, radionuclides are distributed throughout all natural materials, including rocks, soil, plants and animals and their decay is an important component of the ubiquitous radiation 'background'. The sources of radionuclides which are incorporated into natural materials along with their stable analogues are:

(1) Primordial—radionuclides with sufficiently long half-lives to have survived in significant quantities since the creation of the elements which now comprise the Earth. These radionuclides may date back to the origins of the universe ($\sim 10^{10}$ years) or of the solar system ($\sim 5 \times 10^{9}$ years) and include isotopes of U, Th, K and Rb. (Within this category can also be included shorter lived radionuclides continually produced by the decay of long-lived parents—the natural decay series.)

(2) Cosmogenic—radionuclides continually produced in the upper atmosphere by its reaction with 'cosmic rays'. These include ^{3}H, ^{14}C and ^{10}Be.

6

(3) Anthropogenic—radionuclides artificially created and released by man, including both the 'fallout' from nuclear weapons tests and operational releases (primarily in liquid and gaseous form) from commercial nuclear power facilities, hospitals, research, and isotope production. A very wide spectrum of radionuclides is involved here.

Apart from radiation exposure from material containing these radionuclides, direct radiation exposure also results from cosmic rays, medical applications and industrial sources (particularly X-rays in the latter two cases), or simply from certain luminous paints on wristwatch faces.

RADIOLOGICAL PROTECTION PRINCIPLES

Since it was first recognized that exposure to ionizing radiation was potentially harmful, a field known as radiological protection (or sometimes radiation protection, or health physics) has grown up, concerned with quantifying these detrimental effects and setting guidelines to allow the safe handling of radioactive materials. As previously discussed, different radiations interact with matter to varying extents. In addition the chemistry of the various radionuclides (which can occur as solids, gases or liquids), controls their behaviour in the environment and their uptake into food chains and the human body in many different ways. The information now exists to convert measured radionuclide concentrations in environmental samples (soil, water, food, or even building materials) into 'doses', by making various assumptions about potential exposure mechanisms to the material involved (e.g. ingestion, living in contact with them, and so on). These can be related to the health hazard involved. The units of dose and radiation exposure (Sievert, Gray) have already been defined.

The natural background of radiation itself gives rise to a radiation dose which varies greatly with geographical location and the lifestyle of particular

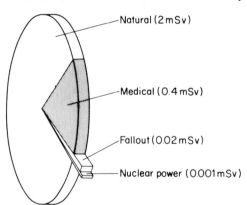

Figure 1.1 Average annual effective dose equivalents from natural and man-made sources of radiation. After UNEP (1985)

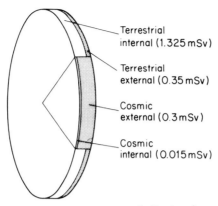

Figure 1.2 Average annual effective dose
equivalent from natural sources alone.
After UNEP (1985)

individuals or population groups. The contributions to the radiation dose of an average member of the world population are shown in Fig.s 1.1 and 1.2. Populations living in areas containing rocks with high abundance of minerals containing U and Th, or at high altitudes where cosmic radiation is less attenuated by the atmosphere, may be exposed to much higher doses. For example in Switzerland the average exposure is ~2.4 mSv but actual values range from 2 to 5 mSv. Additional exposure to radiation results from operations involving active materials. Consequently such operations are governed by radiological protection standards which aim to control doses to individuals.

Radiological protection standards are set by national authorities and differ in detail from one country to another. However, in most countries, specific standards are derived from the same basic principles. These principles are determined by international consensus amongst scientists. The primary body for obtaining this consensus is the International Commission on Radiological Protection (ICRP). The ICRP consists of a main commission and a number of committees comprising experts in various aspects of radiological protection. The recommendations of the ICRP, while not legally binding in any way, form the basis for much of the legislation (both national and international) on radiological protection.

ICRP considers all the situations in which human beings can be exposed to radiation, including natural background, medical irradiation and the generation of electricity by nuclear power. Its recommendations cover both occupational exposure to radiation (i.e. to individuals who routinely work with radiation, for example in the nuclear industry) and the radiation exposure of the public. The central principles of radiological protection are expressed in ICRP's currently recommended system of dose limitation (ICRP, 1977). This system can be summarized as follows:

(a) no practice shall be adopted unless it has a positive net benefit;
(b) all doses shall be kept as low a reasonably achievable (ALARA), economic and social factors being taken into account;
(c) doses to individuals shall not exceed the limits recommended by ICRP.

The underlying philosophy of the ICRP dose limitation system is concerned with limiting the risks of exposure to radiation to acceptable levels. It is important to note the difference between this philosophy and that of reducing risks to zero. ICRP has considered the available evidence on the effects of ionizing radiation and concluded that, for the purpose of radiological protection, it is prudent to assume that all radiation exposures, however low the level, entail some risk. Reducing exposures to zero is both impractical and unnecessary, given natural background radiation, hence the emphasis on acceptable risk.

The three features of the ICRP system of dose limitation are often referred to as 'justification' of practices, 'optimization' of protection for a justified practice and 'compliance with dose limits'. The justification requirement will not be discussed further here because it is not immediately applicable to management of radioactive waste. The generation of electricity by nuclear power inevitably involves the production of wastes which have to be managed, thus it is not necessary to 'justify' waste management as such. Waste management considerations may enter into a justification (in the ICRP sense) of the overall practice of the commercial use of nuclear power. However, while radiological protection aspects will play a part in decisions on energy sources, many other factors have to be taken into account. On the whole it is unlikely that the radiological aspects of waste management will ultimately be an overriding consideration in these decisions, in spite of the attention paid to these aspects in countries such as Austria, Sweden and Switzerland in recent years. The major features of the ICRP recommendations in the context of waste management are therefore optimization and compliance with dose limits. Before discussing optimization, that is the ICRP recommendation that doses should be 'as low as reasonably achievable', it is necessary to consider whether the ICRP system of *dose* limitation can be applied directly to waste management practices. The system clearly is applicable to practices such as the direct discharge of low-level gaseous and liquid wastes into the environment because in these cases it is certain that radiation doses, however small, will be received. In the case of waste disposal options where the intention is to contain wastes for a long period (e.g. by burial in geological formations) the situation is more complex. The risks associated with these options have two major components: the probability that radioactivity will be released into the environment and the probability that the doses received after a release will give rise to harmful effects. Both of these components of risk need to be considered in assessing these options. Neglecting the probability of release would, logically, lead to the rejection of all options in which wastes are contained, because it will always be possible to identify a release situation, even if it has a very low probability of occurring, which would lead to doses in excess of ICRP limits. The application of dose criteria to these options would thus lead to decisions which are inconsistent with those taken about other practices, for example building oil refineries, chemical plants or nuclear reactors. While the possibility of a high consequence accident occurring at such plants is recognized, the low probability of occurrence of accidents is (implicitly in some cases and explicitly in others) also taken into account.

It is obvious from the discussion above that a system of dose limitation, such as that recommended by ICRP, cannot be immediately applied to all waste disposal options. ICRP itself, and other international organizations, have recognized this problem, and it has been discussed extensively over the past few years (see Hill and Webb, 1985). The consensus which emerges from these discussions is that it is necessary to have radiological protection criteria for solid waste disposal options which are framed in terms of *risk*, where risk is defined as a combination of the probability of occurrence of releases of radionuclides into the environment, and the probability that the subsequent radiation doses will lead to harmful effects (ICRP, 1985; NEA, 1984b). The basic form of these criteria consists of:

(a) a requirement that the risk to any individual, at any time, shall not exceed a specified limit
(b) a requirement that the total risk to populations shall be as low as reasonably achievable, economic and social factors being taken into account.

The rationale for such criteria is the same as that for the ICRP system of dose limitation: the limit ensures that no individual is subject to an *unacceptable* risk, while the ALARA principle is designed to ensure that disposal options are compared on risk, cost and other grounds, in order to determine which one is *acceptable*, from a radiological protection point of view. The risk limit recommended by ICRP is 10^{-5} per year; that is a chance in 1 in 100 000 that an individual will suffer a serious health effect.

However, ICRP also recommend that national authorities should select a fraction of this limit for application to a particular waste disposal practice, because individuals could be at risk from other sources of radiation, and it is necessary to allow a margin for unforseen future activities. In the UK, the responsible government departments have set a risk 'target' of one in a million per year for the proposed new land disposal facilities for low and intermediate level wastes (DoE, 1984; see Appendix II).

Let us return now to the ALARA or optimization principle. In the past, there has been much confusion over what it means, and ICRP itself has not been very clear on some points. In particular, there has been a tendency to regard ALARA as an automatic recipe for taking decisions, and to equate it with cost-benefit analysis. Neither of these interpretations is correct, as ICRP and others have, somewhat belatedly, stated (ICRP, 1985; Webb *et al.*, 1986). In waste management, ALARA applies at four levels:

(a) comparison of design alternatives for a specific facility (such as a geological repository);

(b) comparison of different disposal options for particular wastes (e.g. land disposal versus sea disposal);

(c) comparison of different overall management systems for particular wastes (e.g. direct discharge as an effluent versus trapping, immobilization, and disposal as a solid);

(d) comparison of complete waste management systems, including conditioning, storage, transport and disposal alternatives for a given source (e.g. a reactor) or practice (e.g. nuclear power).

At each of these levels, ALARA may need to be implemented in different ways, because the range of pertinent factors involved varies. The decision-aiding technique to be used in an ALARA study will also vary from one level of decision to another. For comparison of detailed design alternatives, cost-benefit analysis may well be appropriate. However, when comparing complete waste management systems, techniques which can handle more factors (e.g. decision analysis) are more suitable. Whichever technique is used, it is important to recognize that its main value lies in structuring judgements, and helping to achieve consistency, clarity, and well-defined reasoning. Furthermore, it is obvious that in a final decision on the course of action to be followed, factors other than those related to radiological protection will need to be considered. Thus the results of any ALARA study are just one input to a decision.

While there are a number of potential problems in applying the ALARA principle to waste management, recent work has shown that these are more apparent than real (Hill and Smith, 1986; DoE, 1986). Nevertheless, much remains to be done to implement this principle and other radiological protection criteria for waste management, at both national and international levels.

SCOPE OF THE BOOK

The objective of geological disposal of radioactive waste is to remove it from man's environment and ensure than any releases, remain within accepted limits. This book concentrates principally on disposal of long-lived radioactive waste in deep repositories, for which containment must last for very long periods of time. Much of what will be said can, however, apply in principle to the disposal of other types of waste. For example, the assessment of groundwater flow around a disposal site, the measurement of chemical retardation properties of rocks and soils and the principles of modelling and safety assessment vary only in detail depending on whether one considers a shallow trench for low level waste (LLW), a deep repository for high level waste (HLW), or even disposal of toxic non-radioactive chemicals. Consequently the reader will be able to extrapolate the arguments presented for deep disposal to other situations. Separate consideration is given to shallow land disposal in Chapter 9. Naturally, emplacement concepts and engineering designs are different and reflect the varying requirements for handling and emplacing lower specific activity wastes. Because of the more sophisticated engineering required in a HLW repository, the cost: volume ratio of HLW disposal is much greater than for other waste types. We have used it here as a central example, since it illustrates well the techniques and methodology underlying the safe disposal of all radioactive wastes.

Much of the work presented is based on the UK waste management programmes over the last decade or so, reflecting our own experience. The programmes in Sweden and Switzerland are also extensively used as examples as both have been well documented recently (KBS, 1983; Nagra, 1985 respectively) and they deal with the two types of HLW—unreprocessed spent fuel and vitrified high-level waste from reprocessed fuel. However we also make frequent reference to parallel work elsewhere and to the role played by international agencies in furthering both research, and the development of standards and guidelines for waste disposal.

CHAPTER 2

Radioactive Waste and the Nuclear Fuel Cycle

THE NUCLEAR FUEL CYCLE

Soon after the discovery of the phenomenon of radioactivity, it was noted that nuclear transformations were often associated with the release of considerable amounts of energy and it was speculated that this could be harnessed as a source of power. Although the energy associated with, for example, α-decay can be harnessed on a small scale (a modern example is the heart pacemaker), the fission process was identified as being particularly useful for larger scale power production. Although, as described earlier, the process of spontaneous fission is limited by the half-life of the nuclide involved, the fission process produces energetic neutrons which, if they hit an already unstable nucleus, can cause premature fission. As this fission in turn produces neutrons, it is possible to produce a cascade or 'chain reaction' if the concentration of potentially fissionable (fissile) radionuclide is high enough. It may be noted that one other nuclear process, the fusion of light nuclei, has also shown a large potential for chain reaction and hence power production. Although both reactions have been utilized in their crudest forms as the basis of explosive devices, the 'atom' and 'hydrogen' bombs respectively, only the former has currently been demonstrated as a practical power source and this will be considered in the following discussions.

In a practical fission reactor the fuel, usually uranium (the 'heaviest' element naturally occurring in practicable quantities), is held in a core which is designed to ensure the optimum balance of efficiency of reaction with ease of control. Achieving a sustained chain reaction is by no means easy and generally requires enrichment of natural uranium (dominantly ^{238}U) in the more fissile isotope ^{235}U, and very careful core design. In current power reactors the fuel—usually uranium metal or uranium oxide pellets—is encapsulated in metal rods in the core, although more exotic alternatives have been studied (e.g. liquid fuel reactors).

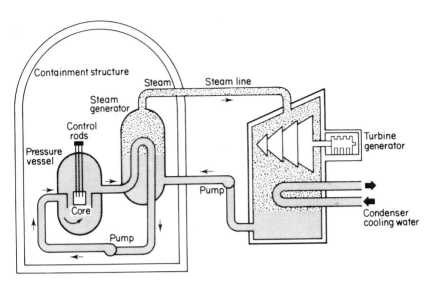

Figure 2.1 Schematic diagram of the cooling circuit of a nuclear reactor—in this case a pressurised water reactor. The primary pressurised water circuit extracts heat directly from the reactor core and passes this on to a separate steam circuit via a heat exchanger. Steam from the heat exchanger passes through the secondary circuit to the electricity generating turbines. A third circuit cools the condenser system.

The fission process produces a considerable quantity of heat. For example, the energy released by the fission of 1 g of ^{235}U corresponds to about one megawatt day (1 MWd) of electricity; sufficient to keep a single-bar electric radiator burning for three years. This heat is extracted by the reactor's primary cooling circuit (Fig. 2.1) and transferred to a secondary circuit and thence to steam to drive electric turbines. As fission progresses a wide variety of light element isotopes ('fission-products') are produced in the fuel. A range of heavy elements (transuranics, such as plutonium, americium and neptunium, with nuclei heavier than the original uranium in the fuel) are also produced when uranium nuclei 'capture' neutrons, hence increasing their number of nucleons. Eventually these new elements interfere with the efficiency of the reaction since they 'capture' the neutrons which sustain the fission process. The fuel element is said to be 'poisoned' and is withdrawn for replacement by a fresh element. The 'spent' fuel is then stored for cooling in a water filled 'pond' (Fig. 2.2) or in some cases in a gas cooled chamber. A considerable amount of heat and radioactivity are emitted as the very short-lived fission product isotopes present in the spent fuel decay.

During the life of the reactor many of its operating mechanical components and the fluid in the primary cooling circuit are subject to intense radiation. Originally inactive elements in these can be converted to active isotopes by this irradiation. Thus many nuclear power plant operations involve contact with contaminated material; irradiated components may need to be replaced as routine maintenance, liquid coolants need to be filtered and scrubbed of

Figure 2.2 Spent nuclear fuel storage and handling ponds; the Pond 5 facility at Sellafield, UK. Casks of spent fuel from Magnox and AGR power generating stations are stored under water at depths of up to 10 m. The facility can also remove fuel elements from the casks and strip them of their cladding ready for reprocessing, and clean and decontaminate transport casks ready for re-use. *Reproduced by permission of British Nuclear Fuels plc*

contaminants (both those in the reactor circuits and those in the cooling ponds) and various other liquid and gaseous effluents will require filtering and decontamination.

The commercial fuel cycle thus involves:

(a) extraction of ores containing fissile nuclides;
(b) chemical purification of particular elements (usually U or Th);
(c) enrichment of fissile isotopes of the element involved;
(d) fuel rod fabrication;

(e) reactor operation;

(f) spent fuel management involving either (i) direct disposal or (ii) reprocessing.

This cycle is shown diagramatically in Fig. 2.3. Alternative fuel cycles exist, but the particular processes involved are inherently similar to those above, although more 'loops' in the process flow chart may be involved.

One fuel cycle which has particular current relevance to the field of radioactive waste disposal is that of the so-called 'breeder reactor'. Although we have previously explained how ^{235}U is enriched to prepare reactor fuel, both ^{238}U and natural ^{232}Th can be transformed into fissile material in a breeder reactor. The basic principle is that the breeder reactor core is surrounded by a blanket of such non-fissile material, which absorbs excess neutrons to create a fissile product. The conversion of ^{238}U to fissile ^{239}Pu is most efficient using energetic neutrons from the reactor core (so-called 'fast' neutrons), while ^{232}Th transforms most efficiently to fissile ^{232}U with the less energetic 'thermal' neutrons. Thus the 'fast breeder' reactor does not breed fuel quickly, it breeds it slowly with fast neutrons. The breeder reactor has the advantage that its incorporation into a reactor power programme considerably extends the fuel resources for other non-breeder reactors, but an essential component of such a mixed programme is the need to process the activated material from the breeders, and to reprocess their fuel.

Finally, it might be mentioned that the fission products and the transuranics (and their decay products) would be absent from the fuel cycle of any future *fusion reactor*. Thus although the problems of waste management would be fundamentally similar to those for a fission reactor, the main concern would be

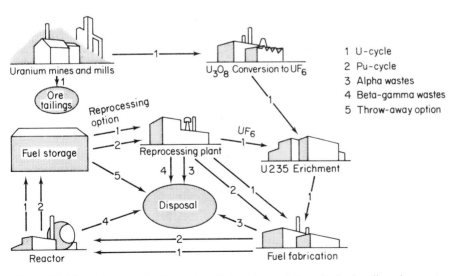

Figure 2.3 The main steps in the nuclear fuel cycle, and the principal radioactive waste streams generated which require disposal. *Reproduced by permission of UKAEA*

large quantities of short-lived radionuclides (such as tritium and activation products), since production of long-lived waste from this source would probably be negligible.

TYPES OF RADIOACTIVE WASTE

Radioactive wastes arise at all stages of the nuclear fuel cycle. In addition, substantial quantities of waste will be produced by the dismantling (decommissioning) of nuclear facilities. Other sources of waste are military uses, and research, medical and industrial applications.

As noted earlier, we are primarily concerned here with long-lived wastes resulting from the civil nuclear fuel cycle. However, the boundaries between particular waste classifications, options considered for disposal of each waste type and even between civil and military nuclear programmes are often rather poorly defined and may vary considerably from country to country. In this section, therefore, an overview of all major waste sources will be presented along with a description of their current classification and the amounts arising, using the United Kingdom as an example.

Each of the steps of the nuclear fuel cycle outlined previously results in waste production. Any activity involving the handling of radioactive materials or their chemical processing produces contaminated waste material. These cover a huge spectrum of solid, liquid and gaseous by-products ranging from slightly contaminated paper tissues through to large items of machinery (lathes, glove boxes, even whole railway engines). If spent fuel is reprocessed to remove the useful uranium and separate the fission product poisons and higher actinide elements (usually at a separate centralized reprocessing plant), then the spent fuel waste form is converted to a 'reprocessed waste'.

The choice of whether to reprocess spent-fuel or regard it as a valueless waste is a very difficult issue, shadowed with considerable political overtones. Reprocessing extracts reusable uranium and valuable isotopes for alternative fuel cycles, and potentially for nuclear weapons. It also produces further large volumes of wastes. The decision in any country is thus based on a host of influential factors, including national energy and defence policies, views on non-proliferation, overall environmental impact and the cost and availability of uranium ore. Consideration of all these issues would comprise a book in itself.

Reprocessing is a chemical operation involving dissolution of the spent fuel in acid and extraction of the uranium and plutonium by precipitation. Consequently reprocessed waste first emerges as a very radioactive and highly acidic liquid concentrate (Fig. 2.4). Reprocessing thus produces its own types of operating wastes, principally:

(a) the metal cladding off the fuel elements (Fig. 2.5);
(b) sludges from corrosion of the cladding during pond storage;
(c) ion-exchange resins from effluent treatment;
(d) medium-activity liquid wastes;

(e) the high-level waste itself;

(f) miscellaneous contaminated laboratory and other materials.

The final source of waste in the nuclear fuel cycle comes from dismantling of the reactor itself. During its operational life, the core of the reactor and surrounding containment and heat transfer structures become radioactive via either activation by the radiation field, or contamination by radionuclides released from the fuel. After a 'cooling period' to allow decay of short lived radionuclides, the reactor is dismantled and its parts separated on the basis of material type and level of radioactivity. The waste is mainly comprised of structural materials (steel, concrete and so on) and concentrates resulting from surface decontamination operations. In general however, following through the sequences of processes listed earlier for the fuel cycle, the volume of waste tends to decrease as its specific activity increases. The nuclear fuel cycle wastes can be considered in very general terms as consisting of:

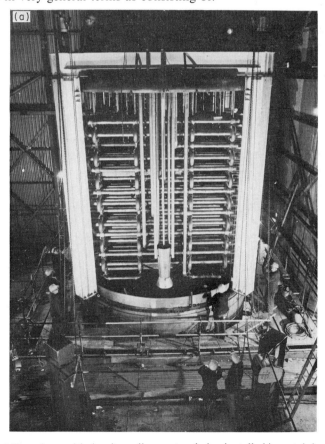

Figure 2.4(a) Complex multi-circuit cooling system being installed in a stainless steel tank for storage of liquid high-level waste generated after reprocessing nuclear fuel. *Reproduced by permission of UKAEA*

18

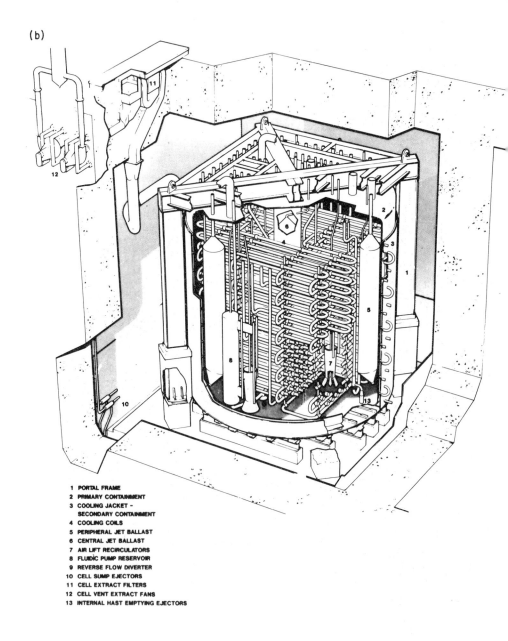

1 PORTAL FRAME
2 PRIMARY CONTAINMENT
3 COOLING JACKET –
   SECONDARY CONTAINMENT
4 COOLING COILS
5 PERIPHERAL JET BALLAST
6 CENTRAL JET BALLAST
7 AIR LIFT RECIRCULATORS
8 FLUIDIC PUMP RESERVOIR
9 REVERSE FLOW DIVERTER
10 CELL SUMP EJECTORS
11 CELL EXTRACT FILTERS
12 CELL VENT EXTRACT FANS
13 INTERNAL HAST EMPTYING EJECTORS

Figure 2.4(b) Diagram of the internal equipment of a liquid high-level waste storage tank. *Reproduced by permission of British Nuclear Fuels plc*

Figure 2.5 Metallic swarf stripped off Magnox fuel elements as a first step in reprocessing. Such fuel cladding debris can constitute one of the highest activity intermediate level waste streams. *Reproduced by permission of UKAEA*

(a) very high volume/low activity wastes resulting from mining and ore processing;

(b) high volume/low–medium activity wastes from enrichment and fuel fabrication;

(c) high volume/low activity wastes from reactor operations, reprocessing and final nuclear plant dismantling (decommissioning) (e.g. waste solutions, gaseous releases, structural materials);

(d) medium volume/medium activity wastes from reactor operation or reprocessing (e.g. ion-exchange resins, sludges and precipitates, plutonium contaminated materials—PCM);

(e) low volume/high activity waste—spent fuel rods or solidified high level waste (HLW) from reprocessing.

Generally speaking the low-activity wastes are also characterized by radionuclides of short half-life, and could also be termed short-lived wastes. While they may contain some very long-lived radionuclides, the concentrations of these are very low, and indeed this is a criterion frequently adopted in national waste classifications.

The high-activity wastes, in addition to many of the short-lived radionuclides, contain high concentrations of long half-life radionuclides. These are the so-called long-lived wastes with which we are principally concerned here, and fall largely into categories (d) and (e) above.

Apart from the nuclear power fuel cycle, two main additional sources of radioactive waste can be indentified:

(a) military nuclear activities;

(b) use of radionuclides in medicine, industry and research.

While naturally occurring radionuclides are certainly mobilized from rocks and released to the environment by non-nuclear industries (e.g. coal-fired power stations, phosphate fertilizer production and so on), consideration of these lies outwith the scope of this book.

The main source of military waste is the manufacture of nuclear weapons. Nuclear reactors are used to produce the 'explosive' from the fuel itself (for fission devices) or from irradiated targets (for fusion bombs).

A further type of predominantly military waste arises from the reactors of nuclear powered vessels (e.g. submarines). When the vessel or the reactor has reached the end of its useful life then the fuel is removed and treated in the same manner as power reactor fuel. The reactor core and other irradiated components (essentially decommissioning wastes on a small scale) must be dismantled and packaged for disposal.

In general, therefore, the waste types are very similar to those derived from a civilian power reactor fuel cycle with reprocessing, although the actual radionuclide inventories in particular waste types may be quite different. The treatment of such waste varies from country to country. In the USA, for example, 'defense waste' is handled completely separately from commercial waste whereas in the UK or France less physical or operational distinction is made between military and commercial wastes. For security reasons, exact details of military waste production are not published. Since their handling and disposal requirements are identical to commercial wastes, and they are subject to the same regulations, military wastes will not be given further specific consideration here.

Radioactive materials are widely used outwith the nuclear power programme. In medicine, for example, radiotherapy utilises the destructive properties of radiation to destroy malignant tissues (cancers) with high specificity, and extensive use is made of radioactive materials for diagnostic purposes. In research, the characteristic emissions from different radionuclides find many applications as 'tracers' or 'indicators' in fields as disparate as biochemistry, chemistry, physics, biology, botany and hydrogeology. In industry and commerce radionuclides can be found in luminous paints, smoke detectors, food sterilisation equipment, batteries for extreme environments (from satellites to heart pacemakers), liquid level meters, radiography rigs, and so on.

Wastes arise from the manufacture, use and disposal of these materials. The total inventory of such wastes in terms of activity is low, relative to those from the fuel cycle, but they are characterized by extremely diverse physical and chemical forms and radionuclide contents. Apart from the very lowest levels of contaminated material, which are disposed of with domestic wastes since they present a trivial hazard, the majority of these research, medical and industrial wastes are shipped back to nuclear establishments for disposal with fuel cycle wastes.

WASTE CONDITIONING

It is clearly much easier to handle many types of waste if they can be turned into some convenient solid package. For some of the very large volume low-activity wastes, which are effectively contaminated trash of widely varying origins, conventional processes of compaction and incineration can be used. The latter technique may give rise to gaseous or particulate radioactive material which either needs to be filtered or subjected to controlled release. The subject of operational gaseous and liquid waste releases is however, outside the scope of this book. Compacted low-activity wastes, together with uncompacted materials are often contained in simple steel drums (Fig. 2.6). The higher the waste activity and the longer its life, the more effort is put into conditioning it with both its intermediate storage and eventual disposal route in mind. Many types of intermediate level wastes are encapsulated in cement, resin, polymers,or bitumen blocks, usually by reducing fragment size by some mechanical means, mixing with a matrix material, and then placing the waste in a steel or concrete container to solidify. (Fig. 2.7). Some waste packages, particularly the concrete forms, may be largely self-shielded, with very low external radiation levels. For some of the higher activity wastes, such as fuel element cladding debris, a process of hot isostatic pressing has been proposed to reduce them to solid blocks of metal integrated with an outer metal container (Fig. 2.8). The processing and

Figure 2.6 Typical drum containing low-level radioactive waste; assorted trash such as might be generated in a laboratory, some of which is contaminated, some of which may only be suspected to be active. *Reproduced by permission of UKAEA*

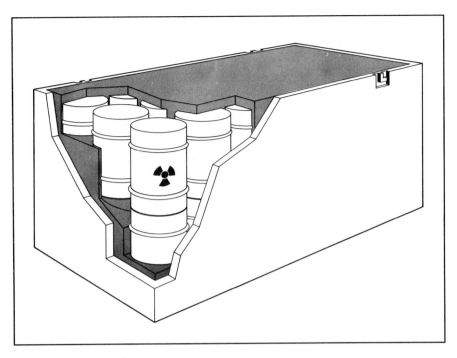

Figure 2.7 Typical solidification system for low or intermediate level waste. Conditioned waste, perhaps in a cement matrix, is contained in metal drums emplaced in a concrete box, backfilled with concrete. *Reproduced by permission of Nagra, Switzerland*

Figure 2.8 Hot isostatically pressed ('hipped') intermediate level waste in a bellows container. In this sectioned sample the container is filled with metallic fuel element cladding fragments, which have been compacted to form a void-free solid. *Reproduced by permission of AAEC/ANU*

conditioning of all types of wastes has been the subject of extensive research and development, and a considerable literature exists, covering many techniques. Of particular interest are the methods proposed, or in use, for the conversion of the most highly radioactive wastes from fuel processing, from their liquid state into a stable and durable solid form, and for their subsequent packaging in a container suitable for disposal.

As noted above, the HLW from reprocessing arises as a liquid, and it is initially stored in this form in specially designed tanks. While storage in liquid form is acceptable for short time periods (decades), it is clear that for longer term storage and disposal it will be necessary to convert the liquid to a solid. At an early stage in research into HLW management it was suggested that liquid HLW could be directly disposed of by injection into rock formations, but it has since been concluded that this procedure would be very unlikely to be acceptable, both because it would be extremely difficult to carry out disposal in this way, and because prediction of the subsequent fate of the waste would be very complex.

Many methods have been proposed for solidification of liquid HLW. The objective is to bind all of the radionuclides into the microstructure of the solid waste product on a molecular scale. The most stable waste form would thus be one in which each waste element was chemically bonded in the atomic structure of a solid which was itself essentially stable and inert. In practice this is not simple as the radioactive decay of the waste isotopes can modify the structure of the waste matrix, and the range of elements and their chemical properties is very large. Some, indeed, are volatile and will tend to escape during the solidification process and thus require separate conditioning. Borosilicate glass (Fig. 2.9) is the most comprehensively studied HLW matrix at present, and is very flexible in terms of waste content. It is however structurally amorphous and is not stable in the presence of high temperature (> 200 °C) aqueous fluids although, as will be discussed later, this is not necessarily a drawback.

Various crystalline and ceramic waste forms have been proposed and many are being tested at present. Mineral assemblages can be chosen which are chemically very stable under all conditions likely to be encountered in a disposal repository. In particular the flexibility of various aluminate and titanate matrix types to variable waste compositions and loadings is being investigated.

Intrinsically these materials are much more stable to leaching than borosilicate glass since the waste elements which can be successfully incorporated are strongly bound in the lattice structures of the host minerals on an atomic scale. Crystalline/ceramic waste forms were first proposed by Hatch (1953) and subsequently investigated principally by McCarthy (1976) in the USA. In 1978 Ringwood proposed a silicate waste form (SYNROC) which he later (Ringwood et al., 1979) modified to a more stable titanate version (SYNROC B: Fig. 2.10). Research on more developed and 'waste specific' types of SYNROC continues, principally in Australia, the USA and UK. Many other 'alternative waste forms' (alternative to glass) are under investigation, including super-calcines, cermets (ceramic beads in a metal matrix) and aluminates (Morgan et al., 1981). The choice of solidification process depends primarily on the properties required for

Figure 2.9 Example of borosilicate high-level waste glass in a steel container. *Reproduced by permission of UKAEA*

the final waste form, in particular its stability and resistance to leaching by water, and on consideration of the risks involved in the solidification process and the ability to carry it out on a large scale. However, the only process which has been demonstrated on a commercial scale is vitrification (i.e. borosilicate glass). Vitrification plants are now operational in France (at La Hague and Marcoule).* In Belgium the Eurochémic PAMELA process will soon be operating, and in the UK the current intention is to build a similar plant. Similarly in other European countries and in Japan, vitrification is the preferred conditioning option.

Both reprocessed HLW and unreprocessed spent fuel elements will require containers in which they can be conditioned, and others in which they can be

* The French vitrification process is known as AVM (Atelier Vitrification de Marcoule) or AVH (de La Hague).

Figure 2.10(a) Bellows container of SYNROC high-level waste form, prior to and after 'hipping'; see also Figure 2.8. *Reproduced by permission of AAEC/ANU*

transported from packaging plant to storage depot and eventually to disposal site. Both during long-term storage and in the disposal facility, additional metal or concrete containers may be placed around the packages in which they were conditioned, and these will be discussed in Chapter 5.

WASTE INVENTORIES

The quantity of waste arising from the nuclear fuel cycle is primarily a function of the power generated and whether or not fuel is recycled via reprocessing. Only at a secondary level are inventories dependant on reactor type and mode of operation. The magnitude of present installed nuclear capacity and projections until the year 2000 are summarised in Table 2.1 (NEA, 1985b).

The first source of waste in the fuel cycle is the process of extraction of the uranium ore. Each gigawatt of electrical capacity of a nuclear plant, denoted by 1 GW(e), requires about 150 tonnes of natural uranium per year which, for an average ore content of 0.2 per cent, produces $\sim 25,000$ tonnes of mining waste. The amount of such waste per unit power generated will, of course, decrease if

Figure 2.10(b) Cut-away of a stack of bellows containers of SYNROC in a full scale (inactive) mock-up of a HLW disposal canister. The void space in the canister may be filled with lead. *Reproduced by permission of AAEC/ANU*

fuel is recycled. The residues of ore extraction, called tailings, have very large volume and, as most of the natural uranium has been extracted, rather low activity, and tend to be dumped on the surface at the mining site (Fig. 2.11). Problems can arise from some of the radioactive daughters of U (especially ^{226}Ra and ^{222}Rn) which are left behind during ore processing.

The very large quantities of such wastes (about 15 million tonnes/year are produced to run the current power plants in the OECD countries) effectively preclude sophisticated geological disposal, and hence they lie outwith the scope

Table 2.1 Estimates of total and nuclear electrical capacity in OECD countries to 2000 (from NEA, 1985b)

	1983			1984			1985			1990			1995			2000		
	Total	Nu-clear	% Nu-clear	Total	Nu-clear	% Nu-clear	Total	Nu-clear	% Nu-clear	Total	Nu-clear	% Nu-clear	Total	Nu-clear	% Nu-clear	Total	Nu-clear	% Nu-clear
Australia	28.7	0	0	31.2	0	0	33.7	0	0	40.0	0	0	47.7[1]	0	0	54.0[1]	0	0
Austria	13.0	0	0	13.0[2]	0	0	13.0[2]	0	0	15.9	0	0	19.3	0	0	19.9	0	0
Belgium[3]	12.6	3.5	27.8	12.1	3.5	28.9	14.2	5.5	38.7	13.7	5.5	40.1	13.3	5.5	41.4	13.3	5.5	41.4
Canada	88.9	7.6[4]	8.5	96.4	9.5[4]	9.9	95.5	10.0[4]	10.5	100.4	13.4	13.3	111.0	15.8	14.2	124.3	15.8	12.7
Denmark	7.6	0	0	7.9	0	0	8.3	0	0	8.0	0	0	8.4	0	0	9.4	0	0
Finland	11.0	2.2	20.0	11.0	2.3	20.2	11.4	2.3	20.2	12.0	2.3	19.2	13.0	3.3	25.4	14.0	3.3	23.6
France	78.4	27.2	34.7	85.6	33.2	38.8	88.1	35.7	40.5	107.6	55.2	51.3	118.6	66.2	55.8	129.4	77.0	59.5
Germany, FR	87.0	11.1	12.8	91.6	16.1	17.6	92.0[2]	16.4	17.8	92.9	22.7	24.4	92.7	22.7	24.5	94.2	24.3	25.8
Greece	5.8	0	0	6.4	0	0	7.1	0	0	9.2	0	0	12.7	0	0	15.5	0	0
Iceland	0.9	0	0	0.9	0	0	0.9	0	0	1.3	0	0	1.7	0	0	2.2	0	0
Ireland	3.2	0	0	3.1	0	0	3.3	0	0	3.7	0	0	3.7	0	0	4.1	0	0
Italy	51.5	1.3	2.5	54.0	1.3	2.4	58.1[2]	1.3	2.2	64.4	3.3	5.1	75.0	6.1	8.1	85.6[2]	12.8	15.0
Japan[5]	139.0	19.1[1]	13.7	142.5[2]	21.8[6]	15.3	148.0	23.7	15.5	170.0	32.6	19.2	197.0	46.1	23.4	223.0	59.5	26.7
Luxembourg	0.2	0	0	0.2	0	0	0.2	0	0	0.2	0	0	0.2	0	0	0.2	0	0
Netherlands	13.9	0.5	3.6	13.5	0.5	3.7	13.5	0.5	3.7	14.5	0.5	3.4	13.6	1.5	11.0	13.1	2.5[7]	19.1
New Zealand[1]	6.6	0	0	6.6	0	0	7.7	0	0	7.9	0	0	8.5	0	0	9.1	0	0
Norway	22.6	0	0	22.8	0	0	24.2	0	0	25.0	0	0	27.0	0	0	29.0	0	0
Portugal	5.6	0	0	5.6	0	0	6.2	0	0	8.2	0	0	9.7	0	0	12.4	0[7]	0
Spain	33.9	3.7	12.0	37.0	4.6	12.4	38.9	5.5	14.1	44.1	7.4	16.8	46.5[2]	9.2[2]	19.8	49.0	10.4	21.2
Sweden	30.95	7.3	23.6	30.95	7.3	23.6	33.05	9.4	28.4	33.05	9.4	28.4	33.65	9.4	27.9	33.65	9.4	27.9
Switzerland	14.2	1.9	13.4	14.7	2.9	19.7	15.2	2.9	19.1	15.3	2.9	19.0	15.3	2.9	19.0	17.7	3.9	22.0
Turkey	6.5	0	0	7.7	0	0	7.9	0	0	14.9	0	0	28.4	1.6	5.6	41.0	2.8	6.8
U. Kingdom	63.7	6.5	10.2	63.7	6.5	10.2	64.0[2]	10.0[2]	15.6	65.6[8]	12.2[8]	18.3	66.3[2]	11.2[2]	16.9	67.0[8]	18.0[8]	26.9
United States	646.2	64.4	10.0	665.4	71.1	10.7	685.5	80.5	11.7	745.8	109.6	14.7	791.6	116.8	14.8	908.0[1]	122.7	13.5
OECD Total (rounded)	1374	158[9]	11.4	1424	182[9]	12.8	1470	204	13.9	1614	277	17.2	1755	318	18.1	1969	368	18.7

[1] Data from 1984 Brown Book.
[2] Secretariat's estimate.
[3] Chooz B1, B2 excluded.
[7] Low projection.
[8] Low scenario numbers selected by the Secretariat from the UK Energy Department's Projections.
[4] Including Pickering 1 and 2 which are down for retubing.
[5] Converted to net figures.
[6] Data from non-governmental sources.

Figure 2.11 Rehabilitated uranium mines tailing heap at Rum Jungle, Northern Territory, Australia. The tailings have been given a semi-permeable clay cap to limit the rate of water ingress and oxidation of the heap, and hence minimize leach rates. *Reproduced by permission of AAEC*

of this book. The environmental impact of uranium mining has been studied in detail in several countries. As examples, the reader is referred to the results of the Ranger Enquiry in Australia (Fox, 1977), and to the large programme of remedial action on tailings heaps in the USA (Matthews, 1986). The latter (termed UMTRA—Uranium Mill Tailings Remedial Action) started in 1983 as a result of federal legislation, and currently involves clean-up and stabilization operations at 24 sites.

The inventory of wastes arising from the other stages of the fuel cycle can be calculated from data provided in the recent Swiss 'Project Gewähr' analysis (NAGRA, 1985). The wastes produced for a nuclear programme of 6 GW(e) with reactor life of 40 years are listed in Table 2.2. From Table 2.1 it can be seen that a similar waste inventory would be expected to be derived from the nuclear power produced each year between 1985 and 1990 by the OECD countries. The production of some types of waste can occur at varying lengths of time after power production, for example if fuel rods are allowed to cool in storage for a certain period of time before reprocessing, and if reprocessing wastes are further cooled before solidification. The present rate of production of wastes in a form suitable for disposal is thus somewhat less than predicted from power production alone. This is illustrated by arisings in the UK calculated by Duncan and Brown

Table 2.2 Wastes arising from a projected 240 GW-year nuclear power programme in Switzerland

	Alpha activity (Bq)	Beta activity (Bq)	Volume (m³)
(a) Reprocessing waste			
Vitrified HLW	3.5×10^{17}	4.1×10^{19}	1 120
Precipitates & concentrates	6.0×10^{14}	1.5×10^{17}	4 320
Ion exchange resins	1.1×10^{11}	1.1×10^{16}	860
Hulls & end caps	1.3×10^{15}	2.9×10^{17}	5 600
Low α technological wastes	1.7×10^{14}	5.3×10^{15}	27 800
High α technological wastes	2.1×10^{15}	5.2×10^{16}	13 900
(b) Operational waste			
Ion exchange resins	2.3×10^{11}	2.4×10^{17}	35 100
Concentrates/slurries	6.0×10^{7}	8.7×10^{13}	1 150
Filters	1.4×10^{10}	1.7×10^{14}	100
Air ventilation filters	4.3×10^{3}	6.8×10^{11}	150
Non-incinerable solids	2.9×10^{7}	3.9×10^{13}	5 430
Incinerated waste	1.6×10^{8}	7.8×10^{13}	560
Fuel element casings	2.1×10^{11}	5.7×10^{16}	1 750
(c) Decommissioning waste			
ILW—activated steel	5.5×10^{9}	1.8×10^{19}	9 480
ILW—activated concrete	—	1.4×10^{14}	1 440
LLW—activated steel	3.5×10^{9}	8.5×10^{14}	5 910
LLW—activated concrete	—	2.7×10^{14}	10 010
ILW—contaminated steel	4.0×10^{10}	3.8×10^{13}	10 220
LLW-contaminated steel	1.7×10^{9}	3.5×10^{12}	38 870
LLW—assorted contaminated materials	2.8×10^{9}	2.8×10^{12}	7 510
Secondary wastes (resins, concentrates, etc.)	3.9×10^{12}	3.9×10^{15}	12 660

(1982) for a power programme of 254 GW(e) years between 1981 and 2000 (Tables 2.3–2.5). These values tacitly include a component from military use of nuclear materials in Britain. More up to date, but less detailed information (DOE, 1986) on the UK waste inventory, gives a useful indication of the variation of waste arisings with different management strategies. The disposal of decommissioning wastes will take on increasing importance by the turn of the century. It is estimated (Lasch, et al; 1984) that in the EEC countries the rate of production of these wastes will equal that of reactor operating wastes by 2050. Almost 2 million tonnes will have been produced by the end of the century, more than 80 per cent of which is classed as LLW, the remainder being ILW.

Waste arising from medicine, research and industry is notoriously difficult to quantify and future production rates are even harder to predict. For any individual country, the quantity of such waste is related to gross national product, degree of technical advancement, and historical background in nuclear research. It is to be expected, therefore, that the rate of production of such wastes will increase with time. To give some idea of types and quantities of waste from

Table 2.3 Wastes arising from commercial power reactors in the UK (from Duncan and Brown, 1982)

Waste type	Untreated waste Specific activity, Ci/m³ α	βγ	Volume to date, m³	Rate of arising, m³/year 1985	1990	1995	2000	Treated waste Volume change factor	Specific activity, Ci/m³ α	βγ	Volume to year 2000 m³
Ion exchange material:											
Magnox	2×10^{-2}	1.4×10^2	250	52	26	13	—	2	10^{-2}	70	1620
AGR	0	50	8	12	16	16	16	2	0	25	580
PWR (primary coolant)	0	6×10^2	—	—	15	90	150	2	0	3×10^2	1950
PWR (secondary coolant)	0	0.1	—	—	9	54	90	2	0	5×10^{-2}	1170
Sludges and concentrates											
Magnox	2	50	325	20	10	5	—	2	1	25	1090
AGR	0	10^3	12	18	24	24	24	2	0	5×10^2	870
PWR	0	10	—	—	4	24	40	2	0	5	1170
Magnox fuel element debris	0	2	1500	112	56	28	—	0.5	0	4	1316
Graphite debris	0	6	1250	90	90	—	—	1	0	6	2240
Filter cartridges	0	6×10^2	—	—	4	24	40	2.5	0	2.4×10^2	650
Fuel element stringer debris	0	10^4	60	90	120	120	120	Storage for decay pending reactor decommissioning			
Miscellaneous irradiated components	0	10^4	2520	230	145	120	90	Storage for decay pending reactor decommissioning			
Miscellaneous contaminated redundant items	0	$10^{-2}–10^2$	4000	700	650	800	900	1	0	$10^{-2}–10^2$	22 000
Desiccant/catalyst	0	5×10^{-2}	32	48	64	64	64	1	0	5×10^{-2}	1260
Miscellaneous solid wastes	0	10^{-2}	5250	5250	5000	6750	8000	0.4	0	3×10^{-2}	55 000

Table 2.4 Wastes arising from fuel fabrication and reprocessing in the UK (from Duncan and Brown, 1982)

Waste type	Untreated waste					Treated waste			
	Specific activity, Ci/m³		Volume to date m³	Rate of arising, m³/year		Volume change factor	Specific activity, Ci/m³		Volume to year 2000 m³
	α	$\beta\gamma$		1981/1990	1991/2000		α	$\beta\gamma$	
High-active liquor:									
metallic uranium fuel	5×10^2	70×10^4	900	50	50	0.3	17×10^2	23×10^5	500
uranium oxide fuel	13×10^3	29×10^5	included in above	–	110	0.4	33×10^3	73×10^5	440
AGR fuel cladding	60	20×10^2	10	–	180	1	60	20×10^2	1810
PWR fuel cladding	60	16×10^3	100	–	180	1	60	16×10^3	1900
Concentrates	10^2	65×10^2	–	–	120	0.7	1.4×10^2	93×10^2	840
Alpha-contaminated floc	10	0.5	6000	50	50	0.1	10^2	5	700
Magnox fuel cladding wastes	25	30×10^2	7250	350	350	1	25	30×10^2	14 000
Ponds sludges	20	6×10^2	1900	100	100	0.4	50	15×10^2	1560
Plutonium-contaminated materials	10	0 (360β)	5000	750	750	0.25	40	0 ($14 \times 10^2\beta$)	5000
Ion exchange materials	2	50×10^2	60	80	80	1.2	1.7	42×10^2	1990
AGR fuel element graphite	0.1	35	–	–	180	1	0.1	35	1800
Miscellaneous items	0.1	50	5000	250	250	1	0.1	50	10 000
Miscellaneous solid wastes	50×10^{-5}	15×10^{-3}	–	20 000	20 000	1	50×10^{-5}	15×10^{-3}	400 000

Table 2.5 Wastes arising from nuclear and non-nuclear research, medical and industrial uses in the UK (from Duncan and Brown, 1982)

Waste type	Untreated waste Specific activity, Ci/m³ α	Untreated waste Specific activity, Ci/m³ $\beta\gamma$	Volume to date, m³	Rate of arising, m³/year 1981/2000	Volume change factor	Treated waste Specific activity, Ci/m³ α	Treated waste Specific activity, Ci/m³ $\beta\gamma$	Volume to 2000, m³
High-active liquor								
MTR[d]	5	92×10^2	510	8	0.17	30	54×10^3	110
DFR[e]	4.4	88×10^2	230	0	0.08	55	11×10^4	20
PFR[f]	16×10^2	16×10^4	—	28	0.04	40×10^3	40×10^5	20
PFR fuel cladding	10	80×10^2	—	30	1	10	80×10^2	600
PFR sludge	5	5×10^2	—	20	0.4	12	12×10^2	140
Plutonium-contaminated materials	3	0	—	150	0.25	12	0	750
Sludge and ion exchange materials (SGHWR)	0	12	250	20	2	0	6	1300
Miscellaneous solid wastes:								
Harwell	1	5×10^2	10	23	1	1	5×10^2	470
Dounreay	0	2.5×10^2	640	100	1	0	2.5×10^{-2}	2640
Amersham	6×10^{-2}	6 (70β)	—	500	1	6×10^{-2}	6 (70β)	10000
Winfrith								
(irradiated items)	1	10^4	5	4	1	1	10^4	85
(contaminated items)	0	1	100	10	1	0	1	300
(others)	0	9×10^2	-	5	1	0	9×10^2	100
Miscellaneous low-level solid wastes (excluding non-nuclear local disposals)	10^{-3}	10^{-2}	60	4400	0.7	10^{-3}	10^{-2}	60000

this source the Duncan and Brown (1982) predictions for the inventory arising by the year 2000 in the UK (a nation with reactor construction, fuel reprocessing and nuclear weapons facilities) can be used. As an order of magnitude estimate, these inventories might be equated to present annual production rates within the entire OECD.

RADIONUCLIDE CONTENT OF WASTES

So far we have discussed wastes in terms of their origins, activities and life-times, without regard to the actual radionuclides they contain. This topic is of course critically important, since when safety analyses of disposal systems are performed, it is found that a few specific radionuclides are responsible for the bulk of potential releases of activity or of predicted radiation doses. This issue is covered in detail in Chapter 10.

Radionuclides are often divided for convenience into 'fission products' and 'actinides'. These classifications derive from their mechanism of production in a reactor. The fission products are the fragments resulting from the splitting of the fuel element nucleus. The fuel itself is one of the chemical group of elements known as the actinides, as are the products formed by activation of the uranium (e.g. Pu, Am, Cm). In terms of atomic weight, the fission products are light (with mass numbers mainly in the 100–150 range) while the actinides are heavy (mass number > 200 daltons). The terms are often used very loosely to distinguish simply between heavy and light nuclides by including their decay chain daughters within the actinide category and light nuclides formed by activation of structural materials with the fission products.

In very simple terms, the fission product radionuclides are generally short lived and characterised by $\beta\gamma$ activity. They include the vast part of the periodic table of the elements. There are exceptions to these generalizations; for example ^{99}Tc, ^{135}Cs and ^{129}I are some of the fission products with very long half-lives. It is such exceptions which in fact form many of the main problems of waste management, as will be discussed later.

The actinides are much more limited in number but are very significant in waste management terms since they include some of the longest-lived radionuclides, such as ^{237}Np, ^{239}Pu, ^{243}Am and ^{247}Cm, and many are α-emitters (consequently being potentially greater biological hazards).

Low-level wastes characteristically contain only low concentrations of fission products (such as ^{90}Sr, ^{137}Cs, etc.) with very low or zero concentrations of α-emitting actinides. As waste 'level' increases so do both fission product concentrations (and hence bulk specific activity of the waste) and actinide content. Long-lived HLW contains high relative concentrations of both groups of radionuclides.

This very simplified description of radionuclide groupings is enlarged on at various points in subsequent Chapters, in particular when considering safety analysis.

Actual radionuclide contents of typical conditioned LLW, ILW and HLW are

Table 2.6 An inventory of radionuclides assumed to be present, at the time of disposal site closure, in cumulative arisings of general low-level wastes in the UK (after Pinner *et al*, 1984)

Radionuclide	Half-life (y)	Activity (Bq) assumed to be present in wastes
Plutonium isotopes and daughter products		
^{238}Pu	8.77 10^1	4.0 10^{11}
^{239}Pu	2.41 10^4	1.0 10^{12}
^{240}Pu	6.55 10^3	1.0 10^{12}
^{241}Pu	1.44 10^1	2.5 10^{12}
^{242}Pu	3.75 10^5	1.0 10^9
^{241}Am	4.32 10^2	1.8 10^{12}
TOTAL[a]		4.2 10^{12}
Uranium isotopes		
^{234}U	2.45 10^5	8.8 10^{13}
^{235}U	7.04 10^8	3.7 10^{12}
^{238}U	4.47 10^9	7.8 10^{13}
TOTAL		1.7 10^{14}
$\beta\gamma$ *activity*		
^{90}Sr	2.91 10^1	4.7 10^{14}
^{93}Zr	9.50 10^5	3.2 10^{10}
^{93m}Nb	1.36 10^1	3.6 10^9
^{99}Tc	2.13 10^5	1.6 10^{11}
^{106}Ru	1.01	8.6 10^{12}
^{107}Pd	6.50 10^6	8.0 10^8
^{110m}Ag	6.84 10^{-1}	2.3 10^9
^{113m}Cd	1.36 10^1	7.2 10^{10}
^{121m}Sn	5.50 10^1	1.3 10^9
^{123}Sn	3.54 10^{-1}	5.0 10^7
^{126}Sn	1.00 10^5	5.6 10^9
^{125}Sb	2.73	6.4 10^{11}
^{125m}Te	1.59 10^{-1}	1.5 10^{11}
^{127m}Te	2.98 10^{-1}	2.4 10^7
^{129}I	1.57 10^7	3.2 10^8
^{134}Cs	2.06	7.5 10^{12}
^{135}Cs	3.00 10^6	8.0 10^9
^{137}Cs	3.00 10^1	5.3 10^{14}
^{144}Ce	7.80 10^{-1}	3.6 10^{12}
^{147}Pm	2.62	7.3 10^{13}
^{151}Sm	9.00 10^1	2.8 10^{12}
^{152}Eu	1.33 10^1	3.5 10^{10}
^{154}Eu	8.60	7.7 10^{12}
^{155}Eu	4.65	4.6 10^{12}
TOTAL		1.1 10^{15}

[a] The total activity is for the alpha-emitting nuclides only.

Table 2.7   An inventory of the longer-lived radionuclides present in intermediate-level wastes (UK) in the year 2000 (after Hill *et al*, 1981)

1. *Type 1 wastes*			2. *Type 2 wastes*		
Radionuclide	Radioactive half-life (y)	Activity in waste (Bq)	Radionuclide	Radioactive half-life (y)	Activity in waste (Bq)
^{60}Co	5.3	4.51×10^{16}	^{35}S	0.24	2.36×10^{11}
^{63}Ni	1×10^2	1.84×10^{16}	^{60}Co	5.3	3.77×10^{12}
^{79}Se	6.5×10^4	1.30×10^{12}	^{65}Zn	0.67	5.66×10^{11}
^{87}Rb	4.7×10^{10}	7.67×10^7	^{90}Sr	28.6	5.67×10^{14}
^{90}Sr	28.6	1.58×10^{17}	^{106}Ru	1.0	2.59×10^{13}
^{93}Zr	1.5×10^6	1.05×10^{13}	^{125}Sb	2.70	1.01×10^{13}
^{94}Nb	2.0×10^4	4.76×10^8	^{134}Cs	2.06	4.03×10^{14}
^{99}Tc	2.1×10^5	4.86×10^{13}	^{137}Cs	30.1	5.66×10^{15}
^{107}Pd	6.5×10^6	3.18×10^{11}	^{144}Cs	0.78	9.29×10^{12}
^{108m}Ag	1.3×10^2	3.40×10^3	^{154}Eu	8.6	6.01×10^{12}
^{113m}Cd	14.6	2.49×10^{13}	^{155}Eu	4.9	8.95×10^{11}
^{121m}Sn	50	3.56×10^{11}	^{210}Pb	22.3	3.22×10
^{126}Sn	1.0×10^5	1.73×10^{12}	^{226}Ra	1.6×10^3	1.68×10^2
^{129}I	1.6×10^7	1.02×10^{11}	^{228}Ra	5.76	3.64
^{135}Cs	2.3×10^6	2.58×10^{12}	^{227}Ac	21.8	1.83×10^5
^{137}Cs	30.1	2.12×10^{17}	^{228}Th	1.91	8.82×10^6
^{147}Sm	1.1×10^{11}	2.96×10^7	^{229}Th	7.34×10^3	1.74×10^2
^{151}Sm	93	8.35×10^{14}	^{230}Th	7.7×10^4	4.41×10^4
^{152}Eu	13	1.10×10^{13}	^{232}Th	1.41×10^{10}	5.3
^{154}Eu	8.6	2.40×10^{15}	^{231}Pa	3.25×10^4	5.49×10^5
^{210}Pb	22.3	5.41×10^3	^{233}U	1.59×10^5	9.58×10^4
^{226}Ra	1.6×10^3	2.83×10^4	^{234}U	2.48×10^5	2.27×10^9
^{228}Ra	5.76	7.55×10^2	^{235}U	7.10×10^8	9.73×10^8
^{227}Ac	21.8	5.31×10^7	^{236}U	2.39×10^7	4.42×10^9
^{228}Th	1.91	1.83×10^9	^{238}U	4.50×10^9	6.19×10^{10}
^{229}Th	7.34×10^3	3.49×10^4	^{237}Np	2.14×10^6	1.02×10^9
^{230}Th	7.7×10^4	7.41×10^6	^{238}Pu	87.8	6.40×10^{12}
^{232}Th	1.41×10^{10}	1.10×10^3	^{239}Pu	2.44×10^6	1.02×10^{13}
^{231}Pa	3.25×10^4	1.59×10^8	^{240}Pu	6.54×10^3	1.48×10^{13}
^{232}U	72	1.79×10^9	^{241}Pu	15.0	6.73×10^{14}
^{233}U	1.59×10^5	1.92×10^7	^{242}Pu	3.87×10^5	1.24×10^{10}
^{234}U	2.48×10^5	6.33×10^{10}	^{241}Am	4.33×10^2	3.93×10^{13}
^{235}U	7.1×10^8	2.82×10^{11}	^{242}Cm	0.45	5.67×10^{10}
^{236}U	2.39×10^7	9.18×10^{11}	^{244}Cm	17.9	1.36×10^{11}
^{238}U	4.5×10^9	1.04×10^{13}	^{245}Cm	8.7×10^3	9.40×10^6
^{237}Np	2.14×10^6	2.05×10^{11}			
^{238}Pu	87.8	8.68×10^{14}			
^{239}Pu	2.44×10^4	2.96×10^{15}			
^{240}Pu	6.54×10^3	3.07×10^{15}			
^{241}Pu	15.0	1.35×10^{17}			
^{242}Pu	3.87×10^5	2.09×10^{12}			
^{244}Pu	8.3×10^7	8.76×10^3			
^{241}Am	4.33×10^2	7.87×10^{15}			
^{242m}Am	1.52×10^2	2.05×10^{13}			
^{243}Am	7.4×10^3	2.50×10^{12}			
^{243}Cm	30	2.72×10^{12}			
^{244}Cm	17.9	2.83×10^{13}			
^{245}Cm	8.7×10^3	1.88×10^9			
^{246}Cm	5.0×10^3	1.08×10^8			
^{247}Cm	1.64×10^7	7.50×10^1			
^{248}Cm	3.7×10^5	5.78×10^1			
^{251}Cf	898	2.01×10^{-1}			

Table 2.8 An inventory of the longer-lived radionuclides present in UK high-level waste, approximately 1000 years after reprocessing (after Hill and Lawson, 1980)

Radionuclide	Radioactive half-life (y)	Activity in waste (Bq)
Fission products		
^{79}Se	6.5×10^4	1.4×10^{14}
^{87}Rb	4.8×10^{10}	8.3×10^9
^{90}Sr	28.8	5.7×10^8
^{93}Zr	1.5×10^6	1.1×10^{15}
^{93m}Nb	13.6	1.1×10^{15}
^{94}Nb	2.0×10^4	4.8×10^{10}
^{99}Tc	2.14×10^5	4.4×10^{15}
^{107}Pd	6.5×10^6	4.2×10^{13}
^{108m}Ag	127	5.8×10^7
^{113m}Cd	14	2.2×10^{-5}
^{121m}Sn	55	5.3×10^7
^{126}Sn	1×10^5	2.0×10^{14}
^{129}I	1.6×10^7	1.1×10^{13}
^{135}Cs	3×10^6	9.7×10^{13}
^{137}Cs	30.17	4.1×10^9
^{147}Sm	1.06×10^{11}	1.1×10^9
^{151}Sm	90	2.4×10^{13}
Actinides		
^{210}Pb	22.3	1.8×10^{10}
^{226}Ra	1.60×10^3	1.9×10^{10}
^{228}Ra	5.76	4.0×10^4
^{227}Ac	21.8	6.8×10^9
^{228}Th	1.91	6.1×10^5
^{229}Th	7.3×10^3	1.9×10^{10}
^{230}Th	8.0×10^4	9.8×10^{10}
^{232}Th	1.41×10^{10}	4.0×10^4
^{231}Pa	3.28×10^4	6.9×10^9
^{232}U	72	5.6×10^5
^{233}U	1.6×10^5	4.3×10^{11}
^{234}U	2.45×10^5	1.1×10^{13}
^{235}U	7.04×10^8	3.6×10^{10}
^{236}U	2.34×10^7	5.9×10^{11}
^{238}U	4.47×10^9	5.8×10^{11}

shown in Tables 2.6–2.8. These are not full inventories of all radionuclides present, but show only those found at significant levels.

THE CLASSIFICATION OF RADIOACTIVE WASTES

By now the reader will have noted that a considerable number of terms have been introduced to describe types of radioactive waste. Many attempts have been made to develop rational and quantitative systems of classification, but the heterogeneous nature of the wastes and their origins can make this very difficult.

It should be noted that all classifications of wastes into a small number of categories are mainly valuable as a means of 'shorthand' reference. The means eventually adopted for disposing of a particular type of waste depend not on what the waste is called, but on its characteristics in terms of radioactivity content and physical and chemical form. This has always been recognized in the UK, and is the reason why the recent study of 'best practicable environmental options' for the management of low- and intermediate level wastes considered each waste stream separately, rather than each category in total (DoE, 1986). In other countries (the USA for example) waste categories are more strongly linked to legislation about their eventual disposal, but in the studies which formed the basis for regulations each waste stream was considered separately. Arguments about whether a particular waste is 'high', 'intermediate', or 'low' level are therefore irrelevant; what matters is that consistent standards and approaches are used in disposing of each waste.

Having said this, there is a widely accepted 'operational' division of active wastes into the so-called low, intermediate and high categories, for shorthand purposes. In the UK the LLW/ILW division is at 4 GBq/t for alpha activity, and 12 GBq/t for beta-gamma activity. The term high-level waste is reserved specifically for reprocessed fuel wastes or spent fuel itself. Thus, for the purposes of this book, the terms LLW-ILW-HLW must inevitably be used rather loosely, particularly when comparing different national programmes.

CHAPTER 3

Principles of Waste Management and Disposal

WASTE MANAGEMENT ALTERNATIVES

The term waste management refers to the complete spectrum of background policy and actual practices which define the classification, control, movement, conditioning, storage and disposal of wastes. The difference between storage and disposal is important to understand. When waste is stored it is *intended* that at some time in the future it will be retrieved, whereas when waste is disposed of it is *abandoned*, with no intention of retrieval. However, in much of the literature storage and disposal are used interchangeably to refer to long term management options. This leads to confusion, not only in research on the technical aspects, but also in the minds of the public and all those involved in waste management decisions. Hence throughout this book the terms storage and disposal will be used in the sense defined above. Views on the need to store and dispose of HLW in particular differ widely. They range from the view that HLW should be disposed of as soon as practicable, because as long as it is stored it presents an unacceptable risk. At the other end of the spectrum there are those who feel that HLW should be stored indefinitely, awaiting the results of research and further technological developments leading to disposal options which are preferable to those currently under consideration. In practice, there are additional complications. Firstly HLW will almost certainly have to be stored for a period prior to disposal (Fig. 3.1 and Beale, 1982) in order to allow the heat output to fall to a level which facilitates handling and also limits thermal effects on the chosen disposal medium. Secondly, in most disposal options it is possible to retrieve the HLW, at least for the period of the disposal operations and in some cases at any time in the future. Retrieval would be very expensive and technologically far more complex than disposal, nevertheless the possibility does exist, and in some countries is, in fact, a legal requirement. There is also a safety aspect to retrieval, and this is discussed again in Chapter 8.

(a)

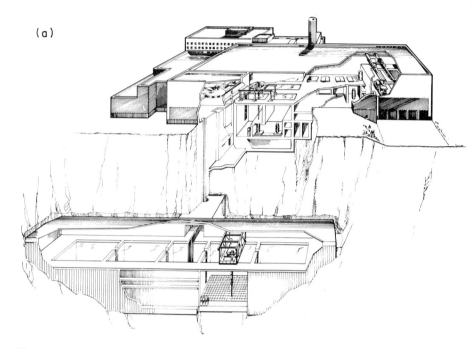

Figure 3.1 Long-term storage of spent fuel and high-level waste. (a) The Swedish CLAB facility at Oskarshamn, for interim storage of spent nuclear fuel in water-filled ponds (similar to those in Figure 2.2) underground in granite, prior to conditioning for eventual disposal (*courtesy of SKB*). (b) The interim air-cooled store for containers of vitrified HLW under construction at Sellafield, UK. *Reproduced by permission of British Nuclear Fuels plc*

Hence the choice of a HLW management strategy is not the simple one of whether to store or dispose. It involves deciding on the method of storage, the method of disposal and the time period for which HLW should be stored prior to disposal. The issues to be considered in the choice are partly technical, partly economic and partly social and political. The debate on storage and disposal extends to many of the lower activity waste types, but it is clear that as waste volume increases the arguments in favour of long-term storage diminish. In reality, long-term storage with no intention to dispose represents an untenable option which merely passes a currently soluble problem onto future generations.

Disposal should thus be seen as representing a complete solution, in that the method chosen should be amenable to predictive analysis of its future performance, to the extent that there can be confidence that once disposed of the wastes will not present any unacceptable hazard at any time in the future. This has important implications for concepts of long-term monitoring of a disposal system. For those systems where extremely long-term containment of wastes is envisaged, it is clearly impractical and unnecessary to attempt to monitor performance.

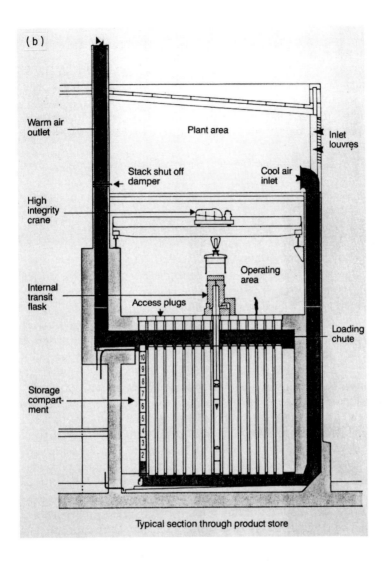

(b)

Warm air outlet

Plant area

Inlet louvres

Stack shut off damper

Cool air inlet

High integrity crane

Operating area

Internal transit flask

Access plugs

Loading chute

Storage compartment

Typical section through product store

BASIC PRINCIPLES OF DISPOSAL

The two fundamental options available for disposal of any material are either to endeavour to keep it in the same place for as long as necessary, or to allow natural processes to mobilize and disperse it harmlessly. The first concept is known as *containment* (or *concentrate and confine*) and is clearly very difficult to ensure for very long periods of time. The second is commonly referred to as *dilute and disperse*. There are obvious attractions to the latter concept for materials which remain toxic for extremely long times. Containment of such materials means that they always remain a potential hazard at one place since their concentration is only slowly reduced by decay. It would clearly be better to design a disposal system which allowed for their mobilization and transport by natural processes, resulting in their dilution in a widely dispersed form at concentration levels which presented no unacceptable hazard.

The attractions of dilute and disperse are naturally counter-balanced by some very difficult problems. If we are to allow the dispersal to take place, then we must be able to model it confidently to ensure that any eventual releases of radioactivity are acceptable. This involves the understanding of a very wide range of processes which can take place during mobilization and transport of the waste.

At present there is an increasingly widespread move towards adopting both concepts in any single waste disposal system. The way this is usually envisaged is that short-lived radionuclides are contained until a sufficient number of half-lives have passed that their concentration in the waste is extremely low. Since containment of very much longer-lived radionuclides for any equivalent number of half-lives is impossible to demonstrate, the system is also designed to allow for their eventual slow mobilization and dispersal. The definition of how long the initial containment period should last depends very much on the waste type and the predicted behaviour of the environment chosen for disposal.

For example, short-lived wastes might be earmarked for shallow-land burial where total containment may be required for around 300 years. After this the total concentration of both short- and long-lived radionuclides in these low-level wastes is very low and almost any degree of dispersal may be acceptable; hence the choice of a shallow rather than a deep disposal environment. In the case of HLW the initial containment period may vary from around 1000 years, up to a million years or more, depending on concept. Some groups suggest that since the dilution and dispersal process in the deep geological environment is so efficient, there are no technical reasons why any initial containment period is necessary, apart from during the operational, pre-closure life of a repository. The basic principles of disposal thus involve the selection of a disposal environment and a system of engineered emplacement of the wastes to provide the required containment and/or dispersal. A general requirement for disposal of any waste is to ensure that the disposal location is remote from man's environment and thus the two methods which have always been chosen historically are burial on land or dumping in the sea. Consequently these two methods have emerged as the principal contenders for disposal of radioactive wastes. It is worth bearing in mind that there have been no adequate controls on disposal of most industrial wastes until very recently (the early 1970s in most countries). This includes the early disposals of radioactive wastes, so there is consequently already an established history of both burial and sea disposal prior to the present day rigorous scientific controls described in this book.

Moving back to a more technical summary of the functional requirements of any disposal system, the following list highlights the main points to be considered:

(a) the extremely long time periods over which isolation is required (indicating dependence on natural as well as man-made barriers);

(b) assurance of low release rates once the period of complete isolation is past;

(c) removal of the wastes from the effects of man's activities or catastrophic natural events;

(d) the technology to implement disposal should be available and the task should be achievable at reasonable cost;

(e) we must be able to model adequately the processes which control the long-term performance of the chosen system, be they physical, chemical or even biological.

How these requirements can be met by various disposal options is discussed briefly below, beginning with geological disposal.

GEOLOGICAL DISPOSAL

Geological formations are, of course, characterized by their great age, and in most environments by their considerable long-term stability. The lengths of time over which certain rock formations have retained their structure, integrity and properties is beyond the scope of human experience, and although the rocks are subject to constant change, the rates are generally extremely slow. This permanence is anecdotal and part of our own folklore: an Icelandic legend has it that there exists in the vastness of the cold northern oceans a huge, smooth dark rock which rises from the ice and the waves. Every year a great black bird goes to this rock to sharpen its beak on the hard surface of the stone. When the bird has eventually worn the rock away then one era of geological time will have passed. Although colourful, this highlights our difficulty in comprehending not only geological time, but by analogy, the isolation periods of long-lived wastes.

Burial of wastes in a suitable geological formation removes them from man's environment and puts them out of reach of major disruptive processes, both natural and man-made. It also provides more than adequate shielding from the radiation emitted by the waste. The wastes are put in an environment where all natural processes which can affect their behaviour are very slow. The deeper the burial, generally speaking, the slower and less pronounced are these processes, and we are effectively matching their rates to the required isolation period of the wastes. Thus it is clear that long-lived wastes will generally require deep disposal, but disposal at shallower depths may be acceptable for shorter-lived wastes. Political expediency also plays a role in choice of disposal depth. Thus some countries have chosen deep disposal for even low-activity wastes because they consider that the extra costs involved are more than matched by the political gain of demonstrable over-design of safety barriers.

The natural geological processes which will affect the behaviour of buried wastes are dominated by the behaviour of groundwater in the host rock unit and in under- and overlying formations. All rocks below the water table are fully saturated with water; that is their pore spaces, which can vary from a few to several tens of percent of the rock volume, are full of water. Even above the water table a considerable proportion of the pore space is water filled. This groundwater moves slowly through the rock, and in doing so it can leach and dissolve the waste and its containers and carry radionuclides away and potentially back to the biosphere. These processes are discussed in detail later. The behaviour of groundwater in the rock (hydrogeology) is thus a very important factor in selecting a host geological environment for disposal, and one is generally looking for very slow and predictable rates of water movement. Many geological environments fit this requirement (see Chapter 6) and a variety of suitable rock types are available for construction of a disposal system, either deep or shallow. Again these are considered in much more detail in later sections.

The disposal systems themselves fall into four main types:

(a) shallow burial in trenches, tumuli, tunnels, concrete bunkers or caissons;
(b) deep injection of liquid wastes into porous strata (with or without a slowly solidifying additive such as cement);
(c) deep burial in a purpose-built or modified existing mine, usually called a repository;
(d) deep burial in boreholes drilled from the surface.

At present the main technique commonly practised is shallow burial of LLW, although some deep injection of low/intermediate level wastes occurs to a limited extent (e.g. IAEA, 1983b). While disposals of more active waste in unconditioned form took place in some countries in the 1940s and 50s (mainly associated with nuclear weapons programmes), the limited information available indicates that such practices have been discontinued.

Geological disposal undoubtedly meets the requirements listed earlier, and in common with other disposal options the main task is in demonstrating that the final requirement of predictability is met.

ALTERNATIVE DISPOSAL OPTIONS FOR LONG-LIVED WASTES

Although geological disposal of radioactive wastes is generally favoured, a number of other options have been studied for long-lived wastes. Generally speaking the high potential costs of these other methods has meant that, with the exception of sea disposal, they have not been considered seriously for low-level wastes. The alternatives for long-lived wastes include:

(a) space disposal;
(b) ice sheet disposal;
(c) disposal beneath the seabed;
(d) rock melting;
(e) nuclear 'incineration'.

Space disposal is one of the most commonly suggested solutions for nuclear waste offered by the 'man in the street'. It has the great apparent attraction of removing waste from the biosphere for all time. Studies have been carried out by NASA which show that the launch of nuclear payloads into a solar orbit halfway between the earth and Venus is feasible, based on space shuttle technology (Fig. 3.2). The main problems involved are the extremely high cost of this option and the risk of launch failure. The disastrous loss of the US shuttle-craft 'Challenger' in January 1986 highlighted this risk, and reinforced the previously estimated rocket launch failure rate of 6 in 100 dramatically. The cost alone means that this option could only be considered for the very highest specific activity wastes, while the launch risks have all but eliminated this disposal route from further consideration. A launch failure could cause widespread, not to say politically embarassing, pollution to almost any region of the world. The repercussions of the $8 million search and clean-up exercise, carried out in the Arctic winter of northern Canada in 1978, following the re-entry of a satellite nuclear power-pack, and its break up in the atmosphere, will long be remembered.

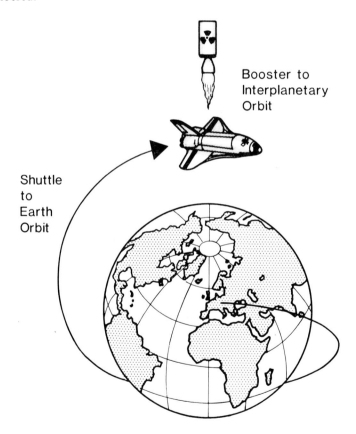

Booster to
Interplanetary
Orbit

Shuttle
to
Earth
Orbit

Figure 3.2 Conceptual space disposal system for high-level wastes which would use the space shuttle and allied technology to deposit waste packages in solar orbit between Earth and Venus

Ice sheet disposal (Fig. 3.3) is another option with great popular appeal. Apart from the remoteness of the polar ice-caps, the fact that heat-emitting wastes would be self-sinking in the ice has clear attractions. The problem areas are the very high transport and handling costs, uncertainties in the evolution of these areas over the geological time spans for which containment must be considered, and legal constraints. At present the controlling limitation is probably the latter, as waste disposal in the Antarctic is expressly forbidden by international law, while alternative sites in Greenland are controlled by Denmark. Disposal of high-level and some types of intermediate-level wastes on the bed of the ocean is prohibited by international agreement (the London Dumping Convention), although it can be shown that the disposal of some wastes by this method presents very low individual risks (NEA, 1985c; Holliday, 1984). In recent years considerable research has been carried out on disposal *beneath* the ocean bed. Advantages of sub-seabed disposal in the deep oceans are physical remoteness, chemical and geological stability of the sediments and rocks, and the great dilution capacity of the sea in the event of releases occurring. In addition, disposal techniques involving self-burying containers ('penetrators') which are allowed to free-fall from the disposal ship (Fig. 3.4), could be relatively inexpensive. Technical problems mainly involve ensuring the performance of

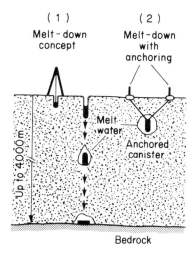

Figure 3.3 Old, and now largely discredited, concepts for disposal of high-level waste below an icecap such as Antarctica or Greenland. (1) Melting concept whereby the wastes own heat eventually brings containers to the base of the ice by melting and (2) a similar concept which prevents the containers from sinking to the bedrock

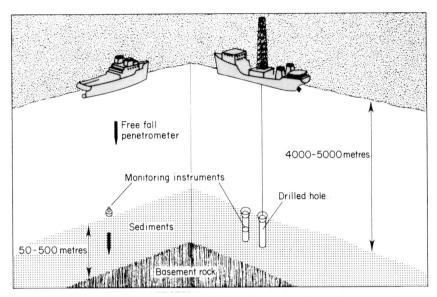

Figure 3.4 Sub-seabed disposal options for high-level waste in the deep ocean basins. Containers of waste free-fall from a ship in a streamlined outer penetrator package and come to rest some tens of metres below the seabed in the stable soft sediments. Alternatively waste packages are emplaced in purpose drilled boreholes in the sediments. *Reproduced by permission of UKAEA*

simple disposal techniques in the deep waters considered, or the high cost and engineering problems with more sophisticated alternatives involving, for example, emplacement in holes drilled in the ocean floor.

Potentially more critical, however, are international concerns. In the forseeable future, pressure from non-nuclear countries with marine seaboards is likely to cause further tightening up of agreements about the use of the oceans for waste disposal, rather than any relaxation. Research and development programmes into sea disposal of long-lived wastes are being reduced, not for technical reasons, but because political factors seem likely to rule out these disposal options.

A more speculative alternative to any of the geological emplacement techniques for high-level waste is 'rock melting'. This involves emplacement of highly concentrated solid or liquid HLW into a deep borehole or cavity. The radiogenic heat produced by the waste melts both the waste itself and the surrounding rock and, upon eventual cooling, the wastes are incorporated into a natural rock matrix. As yet, the technical knowledge required to evaluate fully waste/rock interactions in this system does not exist and hence it has been considered with only very low priority.

The final technique considered—nuclear transmutation, or 'incineration'—is not a disposal option as such, but provides a method for destroying the long-lived radionuclides which present most of the problems in other disposal systems. Very simply, long-lived radionuclides (mainly actinides) are chemically extracted from the waste and emplaced in the neutron flux in a reactor core. This causes them to be activated or fissioned to form shorter lived products. This technique can effectively destroy many of the most potentially harmful nuclides, but is very expensive with presently available technology. In addition, the large amount of processing involves handling vast quantities of very short-lived and hence extremely active radionuclides. This work results in considerable radiation exposure to the work force involved which must be balanced against potential exposures to future populations, which are predicted to be very low (or negligible) in direct disposal techniques. This is an important factor which also has to be taken into account in waste conditioning. Present exposures to an existing work force are real, whereas predicted small scale benefits gained in the very long term, for example by having a more durable waste form, are speculative and lack reality.

CHAPTER 4

Deep Geological Disposal

The previous chapter described many of the options which have been advanced for the disposal of radioactive wastes. Among the geological options for long-lived waste there are only two serious contenders as far as current research effort is concerned; deep-borehole disposal and emplacement in a mined repository. By far the biggest slice of the international research effort has been going into the latter. However, deep borehole disposal (Fig. 4.1) has lately been receiving serious attention in Denmark, Italy, Australia and the USA. Generally speaking mined repositories are projected to be at depths of 300–1500 metres, while deep boreholes can in principle be several kilometres deep.

A mined repository for long-lived wastes may be purpose built (the commonly advanced concept for HLW) or might comprise a modified pre-existing mine or cavern. This latter option is finding increasing favour for the non-heat emitting ILW's, whose exact geometrical disposition in a repository is not crucial from the viewpoint of thermal behaviour. In this case use can be made of any suitable space available. Deep borehole disposal involves lowering waste packages down wide boreholes drilled from the surface, with subsequent sealing of the holes (ELSAM, 1981; ONWI, 1983). In theory boreholes can penetrate many kilometres into the earth's crust where very little water movement is thought to occur. However, this technique is also now being suggested for clays at relatively shallow depths, similar to those of a mined repository (Chapman and Gera, 1985).

When assessing the behaviour of either type of disposal facility the waste emplacement method makes little difference to the type of information which must be gathered, since similar physico-chemical processes occur in each case. In both cases the potential release mechanisms are similar, only the relative magnitudes and significance of the parameters involved varies. For example, leaching of the waste and transport by groundwaters may occur in both cases, but will probably be more significant in shallower environments. The rock and the waste may be subject to higher natural temperatures in a deep borehole, but

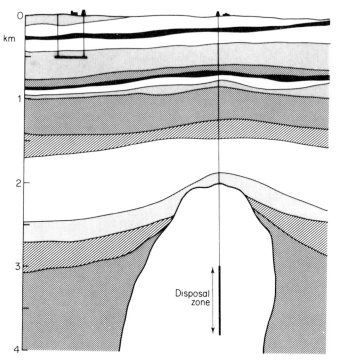

Figure 4.1 Schematic illustration of the deep repository concept for long-lived waste disposal (in this case at a depth of some 500 m in a clay formation) compared to the deep borehole disposal concept (in this case at depths greater than 2000 m in a salt dome)

the mechanisms affecting their behaviour would be similar. Assessment of the performance of either system relies on our ability to model and quantify essentially similar processes. This chapter is a general introduction to the issues involved in deep disposal, while those following examine these processes and the research being carried out to understand and quantify them in more detail, introducing the concept of building the research programme into the safety and practicability of geological disposal around a descriptive model of the performance of a disposal system. In order to do this coherently, a deep repository for long-lived waste is used as a basis for discussion. The wider applicability of the approach is illustrated by a subsequent, less detailed consideration of near-surface disposal of LLW in chapter 9.

MINED REPOSITORIES FOR LONG-LIVED WASTES

Repository disposal of any type of long-lived radioactive waste should enable packages of waste to be taken underground and emplaced in a controlled and repeatable fashion in a structure that can eventually be vacated, backfilled, sealed

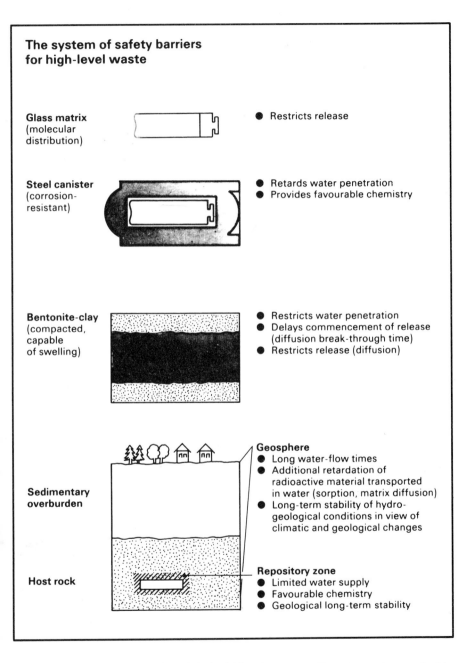

Figure 4.2 Schematic illustration of the multi-barrier system of waste containment; in this instance the Swiss concept for disposal of high-level waste. *Reproduced by permission of Nagra, Switzerland*

and abandoned. A repository design should allow for waste emplacement to take place as a matter of routine over the period in which it operates, and should ensure that this process is as simple as possible within the constraints of safe operating procedures and reliable quality assurance. The repository design is fundamentally intended to ensure adequate long term isolation of the waste and thus comprises a system of natural and engineered barriers to prevent or control the release of the waste from the packages, and the subsequent movement of radionuclides from the waste back to the biosphere. These barriers are designed to counteract or minimize the natural processes which could lead to release of the waste, and in some cases also to minimize the effects of inadvertent human intrusion into the disposal zone.

The idea of having a series of barriers against natural processes has been called the 'multibarrier concept', in which each barrier is nested inside the previous one like a set of Russian dolls. It has been argued that each barrier by itself should be capable of ensuring adequate isolation, in other words that each barrier should be made redundant by the others. However, it is now widely accepted that for long-lived wastes it would be impractical to ensure this function for each barrier, and there is no objective technique for specifying the requisite number of barriers in a system of multiple redundancy. It is more realistic to base safety assessments and comparisons of options on the assumption that the barriers act in concert, and that the behaviour of each can be quantified and predicted. One recent development towards partial redundancy is the suggestion that waste packages can be designed such that the concentrations of radionuclides which can be leached into water in contact with them under any reasonable geological disposal conditions are below accepted limits for drinking water (e.g. Chapman and Flowers, 1986). This effectively decouples waste package behaviour from the geological barrier and means that the packages can be produced in advance of the selection of specific disposal sites, since they would always be compatible with any properly chosen repository location. So far this approach has been considered only for certain cement-based ILW forms. In the majority of conceptual HLW repository designs the multi-barrier aim is achieved by a series of four components (Fig. 4.2). The first is the waste form itself, which immobilizes the waste radionuclides in a solid matrix which is easily manageable

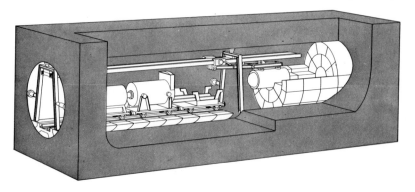

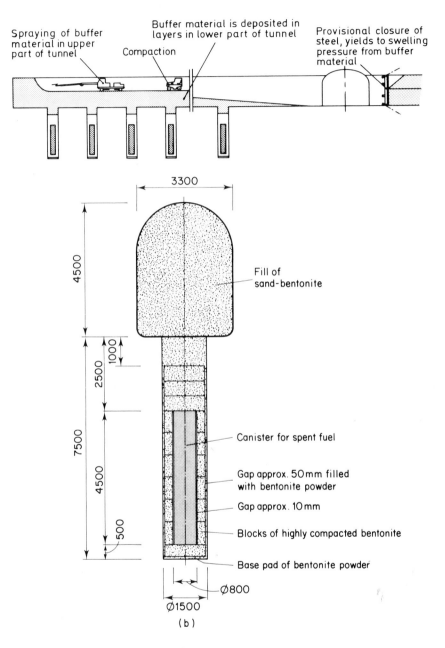

Spraying of buffer material in upper part of tunnel

Buffer material is deposited in layers in lower part of tunnel

Compaction

Provisional closure of steel, yields to swelling pressure from buffer material

3300

4500

7500

1000

2500

4500

500

Fill of sand-bentonite

Canister for spent fuel

Gap approx. 50 mm filled with bentonite powder

Gap approx. 10 mm

Blocks of highly compacted bentonite

Base pad of bentonite powder

∅800

∅1500

(b)

Figure 4.3 Alternative emplacement techniques for high-level waste containers in a deep repository in hard rock. (a) The Swiss concept for in-tunnel disposal. (b) The Swedish concept for disposal in shallow boreholes below the tunnel floor. Both concepts use a buffer material of highly-compacted bentonite around the waste containers, emplaced as preformed blocks (dimensions in mm). *Reproduced by permission of Nagra, Switzerland and SKB, Sweden*

prior to disposal and relatively stable under disposal conditions, and thus provides an immediate barrier which limits the rate of release. The waste will generally be sealed into a metallic canister, and in the case of vitrified HLW, will have been manufactured in such a container. This canister, perhaps enhanced by the addition of a second metallic or ceramic 'overpack', will constitute the second of the barriers. When the containers are emplaced in the rock it is necessary to fill the annulus of the emplacement hole with some form of backfill both to ensure that the waste block is in thermal and mechanical continuity with the rock, and to minimize water movement around the container (Fig. 4.3). This backfill can be designed to perform many physical and chemical functions to help contain the waste, and these are covered in detail in the next chapter. It is generally called the *buffer*, and is the third of the barriers.

The final barrier, and the most important in many respects, is the host rock and geological environment in which the repository is constructed. This isolates the waste from the surface and controls water movement and geochemistry on both large and small scales. The first three barriers are within the engineer's control, and can be tailored to suit a particular concept or geological environment. Containers can be made as thick as required or designed to meet specific handling requirements. Waste forms can be selected at will, and buffers can be of any material, or combination of materials. The waste matrix, container and buffer are termed the '*engineered barriers*'. The rock barrier is, of course, outside the engineer's control and can only be modified to a limited extent, such as by local rock grouting. The principal control on the rock barrier is exercised when the site is selected, which is the reason why so much basic geological research is required to ensure an adequate understanding of the properties of rock types and environments considered potentially suitable for disposal purposes. These aspects are discussed at more length in Chapters 6 and 7.

BASIC REPOSITORY DESIGNS

Many conceptual designs exist for waste repositories, several of which are presented for various key rock types in NEA (1984a). Some designs are exceptionally thorough from the engineers point of view, in that very detailed aspects have been considered (e.g. Burton and Griffin, 1980; NAGRA, 1985). However, until the background research on the requirements of each component of a disposal system has been completed, such sophisticated designs serve largely to show that the technology and methods for repository construction already exist, or can easily be developed. In short, there is in most cases nothing particularly novel to the civil or mining engineer about the type of mined facility that may eventually be required. Given that all repository designs incorporate the multibarrier concept, there are two main approaches from which most designs are simply adaptations. The simplest of the two basic designs is one in which the waste containers are emplaced in tunnels in the rock (see Fig. 4.3). Access to these tunnels is gained by vertical or inclined shafts or adits from the surface. The tunnels may exist at any depth, and may be approached either from directly above or some distance away. The whole repository and its access excavations would be backfilled and sealed when the facility was decommissioned. The second design, which is widely proposed for HLW in particular, is simply an extension of the 'tunnel' repository. In this case the waste canisters are emplaced in wide boreholes which are drilled down from the tunnel floors, instead of being placed in the tunnels themselves (Fig. 4.4). The emplacement holes may be vertical or subvertical, and may be deep enough to contain only one canister or many. There are clearly many options open to the designer, whose first step is to consider the mechanical and physical properties of the host rock and the thermal behaviour of the waste that is to be emplaced in it. Slight variants of these basic designs include underground silos for L/ILW, but more radical alternatives exist such as the Swedish 'WP-cave' which is designed to include a 'hydraulic cage' which isolates the entire repository from the surrounding hydrological system (Boliden WP—Contech AB, 1985).

For a period of several hundreds of years after HLW has been buried it will emit a steadily diminishing but significant quantity of heat (Fig. 4.5). The host rock and the engineered barriers must be capable of allowing this heat to diffuse away without losing their ability to act as barriers to radionuclide release and transport. The thermal anomaly created in a body of rock as a result of emplacing waste canisters of known thermal power in a specified pattern can be calculated. Factors such as maximum temperatures attained in different regions, and thermal gradients occurring throughout the repository and the surrounding rock

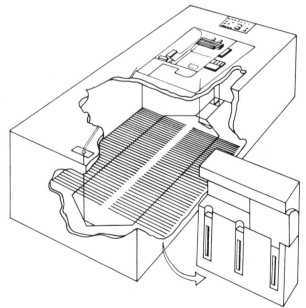

Figure 4.4 Cutaway diagram of the Swedish concept for a high-level waste repository at a depth of about 500 m in hard crystalline basement rocks, using the container emplacement technique shown in Figure 4.3(b) (courtesy of SKB)

56

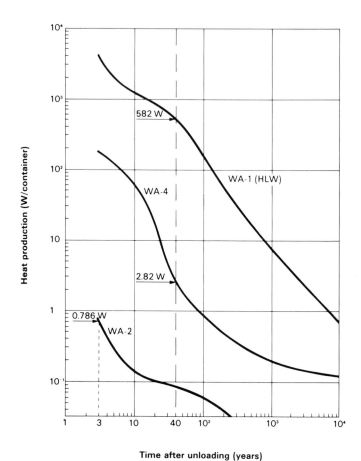

Time after unloading (years)

Figure 4.5 The decrease in radioactive decay heat output of
different waste types as a function of time, using Swiss waste types
as an example. WA–1 is vitrified high-level waste, with an initial
heat output of several kilowatts per container; WA–4 is a higher
activity intermediate level waste (largely fuel element cladding
debris from reprocessing), with an initial heat output of some
hundreds of watts per container, and WA–2 is a lower activity ILW
(precipitates and concentrates) with very low thermal output. The
dashed line at 40 years after production shows the heat outputs per
container of waste at a likely time for disposal. *Reproduced by
permission of Nagra, Switzerland*

can readily be estimated as a function of time. These calculations can be very
precise, since such heat transfer calculations are fairly standard and have been
validated by several in-situ heating experiments (see Chapter 7), and can be used
in the design of a repository. For example the effect of varying the separation of
canisters of waste of given power rating by different distances in a granite
repository is shown in Fig. 4.6. As can be seen, in this case the thermal
calculations show that if individual canisters are placed closer than about 20 m

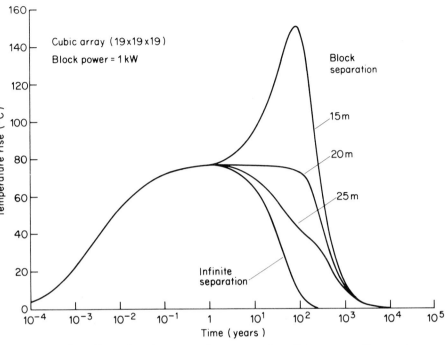

Figure 4.6 The effect of various container separation distances on the maximum temperature rise in the host-rock in a granite HLW repository. If the containers are less than 15 m apart then a sharp rise in temperature occurs after about 100 years. Conversely it can be seen that placing the containers any more than 20 m apart does not give any advantage in terms of maximum repository temperatures. (For a cubic array of 19 × 19 × 19 containers of 1 kW initial heat output at time of disposal; after Hodgkinson, 1977)

apart a very marked rise in rock temperature occurs, whereas there is little benefit in placing containers more than 20 m apart since the maximum temperature rise is the same regardless of increasing distance. Such calculations can thus be used to optimize an emplacement array from the point of view of temperature transients. This type of calculation has been carried out for many rock types and many repository designs, emplacement patterns, container sizes and separations, and initial thermal power ratings of the waste packages considered. The two principal styles of HLW 'tunnel and borehole' repository to come out of such modelling are the 'flat' repository, in which extensive galleries interconnect on the same level, with shallow boreholes penetrating down from them (with perhaps only one waste canister in each hole), and the 'cubic' repository, which utilizes the same type of pattern of horizontal tunnels, although each emplacement borehole may be some hundreds of metres deep and contain many canisters stacked on top of each other with a narrow buffer between each. Other designs envisage more than one level of galleries in a repository (Fig. 4.7).

Thermal calculations, and designs resulting from them, must be carried out in conjunction with other considerations, such as the cost of particular techniques

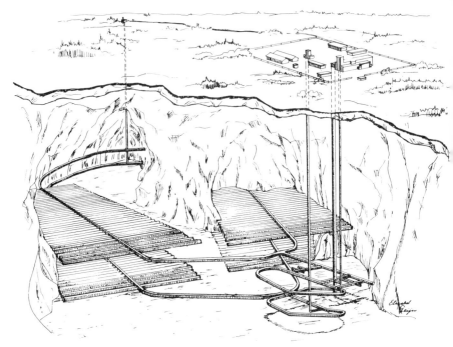

Figure 4.7 Conceptual diagram of a multi-level HLW repository in hard basement rocks (courtesy of SKB)

of tunnelling or boring, the stability of the mined openings and whether they would require any form of support, the ease of transport of men, rock-spoil, machinery, and waste packages in the repository, and for HLW, the means of getting the packages from their radiation shielded transport vessels to the disposal holes in the repository. Repositories in hard rock such as granite or gneiss, and in salt formations, will be self-supporting structures, whereas those proposed for more plastic clays (Fig. 4.8) would need to have their tunnels lined with iron or concrete sections as long as they remain open and operational. Techniques which may be used during the construction and operation of a repository are discussed in Chapter 8.

THE NEAR-FIELD AND THE FAR-FIELD

The preceding discussion has described the physical components common to any disposal system which must be taken into consideration when research is carried out into the events and processes which may break down the integrity of the waste containment barriers. Only when this research has been carried out satisfactorily will more refined designs be feasible.

For practical purposes it is convenient to break up the system of repository barriers into two zones—the near-field and far-field. The *near-field* is the zone which is significantly altered by the presence of the repository while the *far-field* is

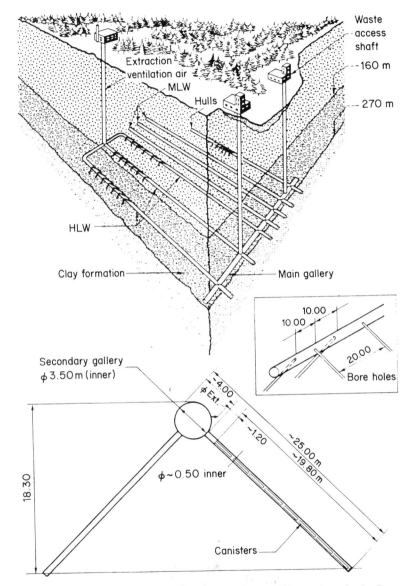

Figure 4.8 The Belgian reference design for a deep HLW repository in the Boom Clay at Mol. Waste containers are placed in angled boreholes drilled from parallel tunnels at a depth of about 270 m (courtesy of CEN/SCK, Mol)

the undisturbed, natural geological system. The near-field includes all engineered barriers plus a region of surrounding rock (usually assumed to extend for some metres or tens of metres) which is significantly altered by heat (for HLW) or chemical releases from the waste package. The near-field is a zone of great chemical complexity in which waste matrix, container backfill and rock all interact with groundwater and each other under the influence of time variable temperature and radiation fields. Overall, this region controls *release* from the repository.

The far-field is very much larger physically and may have quite complex geological structure, but compared to the near-field, it is in a relatively 'steady state' with regard to chemistry, hydrology, and temperature, with any changes subject to normal rates of geological evolution. Overall, this region controls the rate at which water can enter the near-field (thus effectively acting as a massive physical and chemical buffer to the near-field) and also *retards* the transport and *dilutes* the concentration of radionuclides released from the near-field. The output from the far-field goes into the *biosphere* and is the source for calculation of radiation doses to man.

PROCESSES AND EVENTS CAUSING RELEASE FROM THE REPOSITORY

Radionuclides in the waste might escape from a repository by natural causes or as a result of human intervention. The prime objective of disposal of waste is to remove it from the biosphere and man, so one of the aims of burial in a repository is to make it as difficult as possible for future human activities to result in inadvertent exposure to the still-active waste. A backfilled and sealed repository will prevent unintentional access to the waste while some record remains of where a repository is, and what it contains. It is much more difficult, if not impossible, to prevent *intentional* exhumation of the waste. It must however be argued that we cannot be responsible for such action by an informed group at some time in the future. Since such an action would, of necessity, involve a major feat of mining engineering it must be assumed that it would occur only with the full realization of the consequences and risks involved. This type of 'release' of the waste is something which is not taken into account in safety and risk assessments.

What must be taken into account is *accidental* human intrusion into a repository at some time in the future when all records of its existence have been lost. Our experience of myth and legend would suggest that records may endure for thousands of years but this period is very short compared to the half-lives of some important radionuclides in wastes. If technology is still at the same level as at present, or more advanced, it would seem unlikely that any major intrusion by mining activity would occur without advance warning of the presence of such an anomaly in the ground being provided by the type of routine geophysical reconnaissance surveys that generally accompany such operations. The single most significant possibility of future human intrusion is thus considered to result from inadvertent drilling of a borehole through a contaminated zone in the repository, or through a waste container. Some repository designs attempt to prevent such a possibility by emplacing a thick metallic cone above each container to deflect a drill bit. Such features are clearly gestures only, to minimise the chance occurrence of what is in fact, for a deep repository, a very remote possibility. Careful site selection to avoid regions of exploitable natural resources (e.g. aquifers, ore bodies, geothermal zones), is the most important means of reducing the probability of this event occurring.

This is a particularly problematic feature of disposal in evaporite formations. Many evaporite minerals are themselves natural resources, and have been mined in several countries. Such mining may be by remote means, such as solution extraction (pumping water down boreholes to dissolve the salt and return it to the surface in the form of brine) where direct access to the formation is unnecessary. In addition, such formations are increasingly being used for underground storage of oil, gas, and so on, owing to their extremely low permeabilities. This has led some countries to downgrade evaporites as potential host formations, owing to the 'resource sterilization' factor, and the higher probability of future intrusion into a repository.

As far as research into the effectiveness of the containment barriers is concerned, we are thus only interested in considering natural processes which might lead to release of the waste. These fall into two categories: isolated *events* which may either individually or in conjunction cause decreased efficiency or failure of the barriers, and second, the *processes* of slow evolution of the repository and its geological environment which will gradually but inevitably lead to the breakdown of each barrier and eventually result in release of the waste (Table 4.1). This latter category forms the prime consideration in safety assessment, since evolutionary processes are bound to occur and inevitably lead to escape of the waste, and it is onto these processes that the effects of any isolated events from the first category can be superimposed.

An example of the two groups is given by the presence of groundwaters and their movement through a host rock. This will eventually cause the corrosion of the waste canister and the dissolution of the waste. Combined with the slow processes of erosion it will result in some of the waste radionuclides returning to the biosphere, probably at very dilute concentrations, and far into the future. However, a single natural event, such as a massive earthquake resulting in local fault development may 'superimpose' itself on this background pattern of processes, to the extent that a significant direction change or increase of groundwater flow occurs, resulting in earlier or more concentrated return of waste to the biosphere. In order to examine the significance of such sudden events, we must fully understand the background geological processes which they might affect. This has led to the definition of a '*normal-case*' release scenario, which can be modelled deterministically.

In the normal case, release results from the natural processes of geological, climatic and chemical evolution which occur within the Earth's crust. Research focussed on particular geological environments is used to examine whether the normal case leads to releases which are radiologically acceptable or not. The normal case deterministic models, which are used as a basis for designing research programmes, are discussed in more detail in the next section.

Natural events leading to release are related to climate, tectonic patterns and random catastrophes. The last group includes such unpredictable events as the impact of a massive meteorite or asteroid on a repository site which is of sufficient

Table 4.1   Processes and events caused by nature, waste disposal and man (after NAGRA, 1985)

1 *Slow natural processes*
climate changes
sea-level changes
erosion (fluvial and glacial)
sedimentation
tectonic crustal movements
magma intrusion
volcanism
diapirism
diagenesis
metamorphism
weathering, mineralization
groundwater changes

2 *Rapid natural events*
earthquakes
volcanic eruption
meteor impact
flooding with extreme erosion
hurricane, storms
movements at faults
formation of new faults

3 *Those caused by disposal of waste*
radiation damage of the matrix
radiolysis
nuclear criticality
canister movement in backfill
decompressed zones from mining
mechanical canister damage
differing thermal expansion of
● glass matrix and canister
● canister and backfill
● backfill and host rock
● host rock zones
thermal convection
thermally induced chemical changes
drying out and resaturation
chemical changes due to corrosion
geochemical changes in
● backfill
● host rock
physico-chemical processes (e.g. colloid formation)
microbiological processes
gas production
failure of shaft sealing

4 *Caused by man*
direct alterations in hydrogeology
(e.g. through draining or storage)
injection of liquid waste
drilling
geothermal energy production

magnitude to lead to rapid or immediate release of the waste. The probability of such an occurrence can be estimated from our knowledge of such impacts in the geological record and of the origins and distributions of such extraterrestial bodies, but readers should make up their own mind as to the scientific or social validity of considering such esoteric events in an objective safety assessment. Climatic events (rather than processes) can be accounted for in the prediction of evolutionary processes, but in general it is more satisfactory to superimpose the likely effects of, for example, future glaciations or changes in sea level caused by melting of polar ice, onto a normal case groundwater flow model. Thus a probabilistic assessment of events is superimposed on the deterministic model of 'normal case' processes. Increased rates of erosion must be allowed for in upland terrains; caused by glaciers, deepening of river valleys, tropical erosion and so on. However, it would appear reasonable to assume that no individual climatic event is in itself likely to cause release of waste from a deep repository, but will instead have a more or less predictable effect on regional groundwater flow patterns. If such factors are taken into account when designing and siting a repository, then future climatic changes, even those involving marine incursion over the site or submergence beneath an ice-cap, need present no problems of containment. Certain environments would naturally prove unsuitable in such circumstances, and after preliminary consideration would be discarded.

Tectonic processes are not 'sudden', but tend to occur over very long periods of time ($\sim 10^7$ years) and do not in themselves constitute a hazard to the barriers in a repository. However, tectonically active regions of the world are often characterized by volcanic activity, earthquakes and major crustal movements. Regions such as the Mediterranean area, parts of the Middle East, central Africa and the Pacific Ocean margins display such symptoms of continuous tectonic activity, and in these areas difficulties may arise in the prediction of repository performance. Knowledge of the mechanisms and driving forces of these global tectonic processes has increased dramatically over the last twenty years, to the extent that we are able to predict with some confidence which regions of the earth's crust are likely to remain inactive in terms of vulcanicity and frequent major seismic events over the next few millions of years. Clearly, major earthquake damage or exposure to the many paroxismic processes associated with volcanic activity can only be deleterious to the integrity of a disposal facility. This indicates that repositories should be sited away from tectonically active or marginal areas.

The risk of seismic damage is, however, impossible to avoid completely, since all regions of the crust are seismically active to some extent. It is known that underground structures (such as a mined repository) suffer very little damage as a result of even quite major earthquakes, since the host rock and the components of the mine all undergo the same degree of acceleration when a seismic wave passes. The perceived risk lies in the possibility of faults being initiated in the repository rock, or of existing major fractures being reactivated. Once again, it is extremely unlikely that such events in themselves, would lead directly to release or exposure of the waste, but they could lead to modifications in ground-water flow, and accelerate the normal-case process of mobilization and migration. In the case of intrinsically dry host rocks such as salt domes, faulting could result in access of water to the waste from overlying water-bearing strata, hence initiating the slow process of leaching and migration. Since it is impossible to design completely against such events, the safety assessment of a disposal system must take them into account by examining the probability of their occurrence and the effect they would have on release. Information is thus required on the frequency and probability of seismic events of varying magnitudes occurring in specific areas, how these may change with geological evolution, and how they would affect the normal case release model.

Reading through the list of events which might speed up the release of the waste (e.g. Table 4.1) one is struck by the cataclysmic nature of some of the natural processes involved; submergence beneath the oceans or beneath an ice-cap, penetration by volcanic conduits, rupture by massive earthquakes and exhumation by meteorite impact. It is tempting to ask whether, by seriously considering the effects of some of these hazards in the safety assessment, we are becoming obsessed with the containment of radioactive waste when, in the event, the physical damage to the environment would have orders of magnitude greater consequences for mankind. It is for this reason that we have omitted mention of the more extreme results of penetration by nuclear weapons and intentional sabotage during times of war. With such natural or man-made events occurring, a release of radioactive waste would be the least of mankind's worries.

THE CONCEPT OF A 'NORMAL-CASE' RELEASE MODEL

In the preceding section it has been proposed that in any of the major geological environments considered for the siting of a repository, there exist progressive geological processes that must inevitably cause the return of some of the waste to the biosphere. These can be considered using what has come to be called a 'normal-case evolution model'. In fact one basic model can be used to describe almost all environments. The normal-case model (Fig. 4.9) can be constructed around the only credible natural mechanism by which components of the waste can be transported back to the surface, namely transport in groundwaters.

Free water is present in all rocks, and it moves through the pores and fissures of the various lithological units in response to differences in hydraulic 'head' from point to point. Even where no flow occurs, the body of water contained in the pores in the rock can allow diffusion of radionuclides, although at very slow rates. The corroding and leaching action of water will eventually mobilize the waste, in solution or in suspended form, and it will migrate and diffuse with the local and regional movement of the groundwater until it eventually finds its way back to the biosphere. The route which is takes, how long it takes to travel along it, how far and in what form, are the fundamental questions which must be answered before realistic safety assessments can be made.

An exception to this simple model of release of the waste is provided by repositories in dry salt deposits. Since there is little or no interconnected water

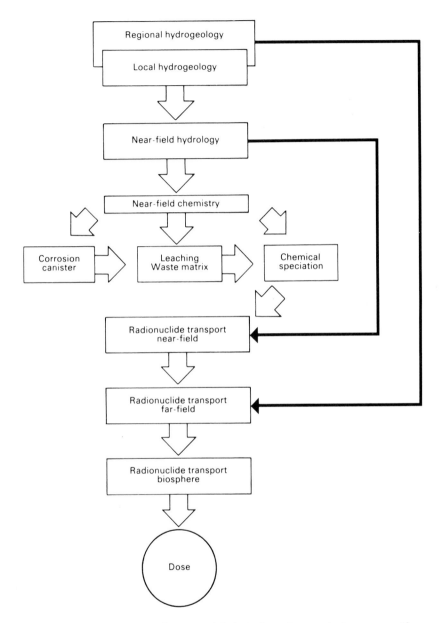

Figure 4.9 A typical normal-case model chain for safety analysis purposes (from Nagra, 1985). *Reproduced by permission of Nagra, Switzerland*

filled porosity in pure evaporite formations, there is no possibility of water flow through these units. (Impurities such as silty or sandy layers can give rise to some component of permeability, and if an intergranular film of water is present, then slow diffusion may be possible, but at rates lower than in free pore waters).

Release should not occur unless some natural event, such as fault development or breach of the access shaft by an overlying aquifer during the operation phase, allows access of water to the waste, or the slow process of diapirism (uplift) moves the salt body *en masse* into a higher water-flow environment. Subsequently, the normal-case model would apply in just the same way. As a consequence, assessments of disposal systems in salt bodies must start by considering the probability of such disruptive events occurring.

The 'normal-case' release model thus starts with water gaining (or regaining) access to the disposal zone of a repository, penetrating the backfill, corroding the overpack and canister, slowly dissolving the waste, mobilising the waste elements in specific chemical forms, and transporting them by long and tortuous paths with various rock-waste interactions occurring en route, until they are eventually discharged into the biosphere. The model takes into account radioactive decay or ingrowth of radionuclides during the time periods involved.

The next two chapters explain how such research programmes have been undertaken, how they are carried out and the type of results being obtained. In keeping with the outlines given in this chapter, the research described is aimed at building a realistic normal-case model for a deep repository for long-lived waste, and assessing the additional effects of other events and processes on this. The process of release starts in the near-field, so the next chapter is dedicated to this region, while Chapter 6 examines the factors controlling nuclide migration in the far-field.

CHAPTER 5

The Near-Field: Radionuclide Mobilization

As defined earlier, the near-field is the zone altered by the presence of the emplaced waste. It comprises all the engineered barriers and extends a small way into the host rock itself. The near-field zone is one of great chemical complexity, influenced by properties of both the waste package and the host rock. The main aim of analysis of the near-field is prediction of the mobilization rates of all significant radionuclides into the far-field, which acts as a '*source-term*' for subsequent migration calculations. Parameter variation in such analysis also allows the relative performance of individual engineered barriers to be assessed, as a first step in system optimization.

In this Chapter the processes occurring in the near-field will be reviewed in very general terms and then the performance of individual barriers will be considered in more detail. Finally, the construction of a source-term model will be explained with some indication of the simplifications required and the advantages and disadvantages of different approaches to this problem.

EVOLUTION OF THE NEAR-FIELD ENVIRONMENT

During operation, a deep repository will be drained and ventilated, ensuring dry conditions while waste emplacement proceeds. After emplacement, the conditions in the near-field will change gradually with time and a number of distinct stages in the evolution of the normal-case model can be recognised:

(1) Water re-invades zones of the host rock which have been drained during operation. The hydrogeological environment begins to stabilise and water starts to penetrate the backfill. For some backfill materials, such as bentonite, this causes them to swell. In confined conditions this builds up a 'swelling pressure' which causes plastic flow to seal any constructional

gaps and, possibly, any disturbed rock zone around the emplacement tunnel.

(2) For high-level wastes, wetting of the backfill from the outside may be coupled to a drying from the inside as radiogenic heat warms the engineered barriers and some region of the surrounding rock. A thermal profile is relatively quickly established, within decades, which slowly decays to insignificance over some thousands of years.

(3) As groundwater penetrates the backfill (or possibly even during transit through surrounding warmed rock) its chemistry alters due to a complex series of reactions including dissolution and precipitation of particular minerals and ion-exchange. If the backfill is simply crushed host rock such alteration may be negligible, but if materials such as bentonite or concrete are used, water chemistry will be completely altered. For HLW, some additional chemical changes (radiolysis) may be caused by neutron and γ radiation which can penetrate the intact canister.

(4) After water has penetrated the backfill it comes into contact with the canister, which begins to corrode. In many designs, the saturated backfill has much lower permeability than the host rock and hence negligible flow occurs in this region and all solute transport (of corrodants in groundwater inwards and corrosion products outwards) occurs by a diffusive process. Such diffusion may be very slow and thus limit canister corrosion rates.

(5) At some point canister failure occurs due to corrosive penetration or mechanical effects (for example due to external hydrostatic pressure or backfill swelling pressure). Even after its failure as a complete barrier to water ingress, remnant canister material and corrosion products may act as a further diffusive barrier and alter the chemistry of water penetrating through the backfill.

(6) When water contacts the waste itself it will begin to degrade and radionuclides will be released, either into solution or as particulates or colloids. Prediction of the chemical environment in this region is very complex, being determined by all the reactions previously considered, plus reaction with the waste form itself (which may be somewhat heterogeneous). For HLW especially, there is the further complication of additional radiolysis from short range α and β particles which can now contact the groundwater. In the low-flow or diffusive environments generally expected, waste degradation is unlikely to proceed by simple dissolution but is likely to involve formation of solid reaction products (secondary minerals) *in situ*.

(7) As the waste form breaks down, radionuclides are dissolved or mobilized as particulates and begin to migrate through the near-field. The rate at which radionuclides are released into solution is generally determined by the breakdown rate of the matrix (although it is possible for some radionuclides to be located preferentially on the waste surface as will be seen later) but may be further constrained by their low solubility. The rate

of migration of radionuclides through the far field is determined by the hydrology of the different rock formations along the transport path and the extent of retardation processes, for example due to sorption onto secondary minerals from the waste, canister corrosion products and backfill. Further constraints may arise from precipitation (or co-precipitation) due to changes in pore-water chemistry along the flow path.

(8) Evaluation of all the stages above, taking into account radioactive decay (or ingrowth), allows prediction of the release of any radionuclide into the far-field to be assessed as a function of time. For some cases, the near-field/far-field boundary can conveniently be considered to lie within the mechanically disturbed rock zone extending at most a few metres from the tunnel or borehole wall. In others chemical perturbations (e.g. oxidizing conditions caused by radiolysis) may penetrate significant distances into the host rock. In the latter case, the development of the chemical perturbation front in terms of solute buffering by the mineralogy of the host rock and radionuclide transport within this region must also be evaluated.

FACTORS AFFECTING THE NEAR-FIELD ENVIRONMENT

In this section we consider the guidelines used to define the requisite performance of the individual engineered barriers and the techniques used to quantify them. Before considering the barriers individually, some parameters dependent on waste type, overall repository design, and location need to be specified.

Temperature

The most important of these factors is probably temperature, especially for HLW. If a repository is constructed at, say, a depth of a kilometre or so, then the ambient temperature will be about 25–50 °C, largely regardless of rock type. The decay heat of emplaced HLW will be additional to this, and the exact temperatures attained in any part of the near-field can be predicted very reliably using heat transfer calculations which take into account the slowly declining heat output of the waste and the thermal conductivities and diffusivities of the materials involved. For more exact calculations it may be possible to include the effects of backfill resaturation to form a coupled heat/water transfer model. Such complex systems are much harder to quantify, and the differences between the results obtained and those from a simpler model are generally small. Similarly, complete backfilling of all tunnels and voids is generally assumed, and heat transfer is thus taken to occur by conduction only, although the small effects of remnant air-gaps can also be evaluated. Thermal convection on a larger scale was once considered to be of potential significance, but a model of thermal buoyancy in fissured rock (Hodgkinson, 1980) has shown that for expected disposal environments, while it may result in large scale water movements, it does not contribute significantly to the transfer of heat. The maximum temperatures attained in any part of the near-field depend principally on the thermal loading of the HLW containers and the geometry in which they are emplaced (e.g. Hodgkinson, 1977; Deane and Hollis 1979). In the rock types considered for disposal there is a predicted rapid increase in rock-edge temperature over the first few years after emplacement, with very steep thermal gradients away from the centrelines of individual waste packages, across the backfill and out into the rock (Fig. 5.1). After between 50–100 years these gradients would start to flatten out such that temperatures within the repository itself would be fairly uniform. This is more obviously the case in the 'cubic' repository than in the 'flat' type (Fig. 5.2). Most countries engaged in repository design exercises have opted for an arbitrary value of 100°C or less for this maximum temperature (e.g. CEC, 1982), although research continues to optimize this figure for the various rock types and configurations involved. In the USA, maximum values of up to 250°C are suggested for some salt and basalt host rocks and 150°C is suggested for tuff. 80–100°C remains the design target for plastic clays. It is likely that any eventual design for fractured cystalline rocks such as granite could tolerate temperatures between 100 and 200°C (Chapman, 1980).

A more serious constraint on the maximum admissible temperature may well be the stability of engineered barriers. Both metal corrosion, and many backfill alteration reactions (e.g. montmorillonite to illite in a bentonite backfill) are

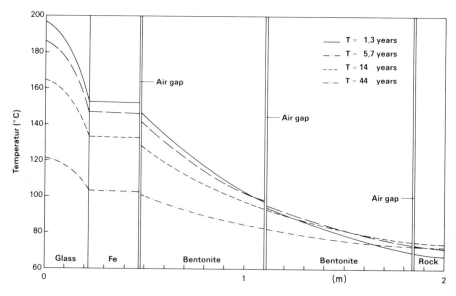

Figure 5.1 Temperature profiles in the near-field of a HLW repository in hard crystalline rocks, based on Nagra (1985). The profiles (at various times after disposal) are calculated for the glass waste form, its iron container, the bentonite buffer in the disposal tunnel, and the host rock. *Reproduced by permission of Nagra, Switzerland*

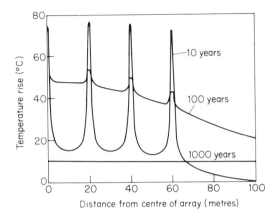

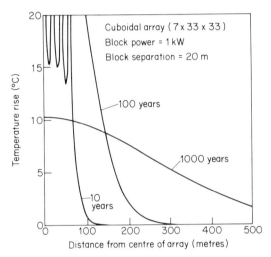

Figure 5.2 Temperature profiles through a complete repository in granite at different times after waste disposal (after Bourke and Hodgkinson, 1977). The calculations are based on a cuboidal array of waste containers of $7 \times 33 \times 33$, with each block of waste having a 1 kW heat output.

much slower at low temperatures. Although, in some cases, the reactions encouraged by high temperatures may be favourable (e.g. formation of protective layers on metals), important properties may be lost; for example, the swelling capacity of bentonite. If long-term pre-disposal storage is considered, then it may well be feasible to keep near-field temperatures below 100°C.

Regardless of the maximum temperatures attained in any part of the repository, the near-field is characterized by steep thermal gradients over a period of several years to some decades depending on precise rock properties and repository design, reaching a maximum temperature after a few decades, and followed by fairly uniform near-field temperatures which decline very slowly over several hundreds of years until they approach what would be natural ambient values during the first millenia. Thereafter, while the actual *amount* of heat given out by the waste is still considerable (integrated over time for the ensuing 10^4–10^5 years it dominates the total thermal energy output; Bourke and Robinson, 1981) its production *rate* is so small that it has little or no effect on processes in the near-field. However, as will be seen in the next chapter, this steady trickle of heat is a major driving force for groundwater movements in some rocks for very considerable periods.

Stress field

A second important factor comprises the hydrostatic and lithostatic stresses. In a hypothetical repository at about 1000 m depth in a fractured or porous rock the water present (assuming complete interconnection of all fissures and pores) in the rock will exert a hydrostatic pressure on the near-field components approximately equivalent to a 1 km head of water. Naturally, pumping during repository operations, recovery of hydraulic pressure profiles after repository closure, and local hydraulic gradients will have some effect on this value, but it represents the minimum stress to which the waste packages would be subjected. In some mixed sedimentary environments deep groundwaters may be at substantially more than hydrostatic pressure, due to transfer of part of the lithostatic load, or long-range artesian effects. Over longer periods of time, as stresses in the rock-mass itself readjust to hole boring and tunnelling and backfilling, the near-field components will be expected to be subject to lithostatic load pressures caused by the one kilometre overburden of rock. These stresses are often anisotropic, particularly in some areas of ancient crystalline rock, and horizontal components of stress can be up to four or more times the vertical values, which are themselves two to three times greater than the maximum hydrostatic pressure, owing to the density of the overlying rock units. Stress readjustment will take place very quickly in some of the plastic clays and salt deposits being considered for disposal.

A third component of the stress field acting on the waste and its packaging is that caused by the heat output of the waste itself. This can cause highly anisotropic stress build-up in response to the steep thermal gradients and differential thermal expansion (or contraction) of rock and near-field materials. Such point stresses tend to congregate around any unconfined surfaces (although these will be uncommon in a backfilled repository) and can cause spalling and microfracturing in the more brittle crystalline rocks. The plastic rocks, such as salt, demonstrate quite rapid creep reponse to thermal stressing, which would tend to close any openings which may be present. The crystalline rocks will also display creep, but on a very much longer time scale, which may be significant in locally modifying the hydraulic properties of the rock. If a repository is designed such that local temperatures are allowed to build up to high values, then such thermomechanical responses must be considered when assessing the stability of mined openings (Blacic, 1981; Hodgkinson and Bourke, 1980). Where thermal stresses occur in a completely backfilled repository they are of considerably less importance.

Hydrogeology

Solute transport in the near-field is influenced by the regional hydrogeology of the host rock formation, the design and construction of the repository and the physical characteristics of the engineered barriers. Even though the regional hydrogeology constrains the maximum water flux through the entire repository, the rate and distribution of solute transport between the near and far-fields depends on the layout of the near-field. To illustrate this, we can consider the case of uniform flow through an homogeneous porous medium. If an emplacement tunnel is drilled through such a formation and backfilled with material of lower hydraulic conductivity than the host rock, water will tend to flow round the tunnel rather than through it (e.g. Fig. 5.3a), and hence the net water flux through the near-field would be less than that in the far-field. The near-field is generally more complex than this, however, and any tunnel will tend to be surrounded by a damaged rock zone of higher hydraulic conductivity caused by the construction process (e.g. Kelsall *et al.*, 1984). In this case, flowlines through the backfilled tunnel would look more like those shown in Fig. 5.3b. In reality, of course, the situation is further complicated by heterogeneities/anisotropies in the properties of both the host-rock and near-field materials. Additionally, in very low-flow environments, molecular diffusion will be the dominant solute transport mechanism and this may need to be coupled to advective transport in more conductive regions.

In order to assess barrier performance, a simpler model of solute transport is often used in which it is assumed that the entire water flux passing through the repository can exchange solute with the near-field, and is shared equally between individual waste packages. This is often termed the 'equivalent flux' or Q_{eq} (e.g. KBS, 1983) and expressed in litres/container/year, as if all solute transport occurs by advection—even in a diffusion dominated environment. The extent of conservatism in this approach can be evaluated by more complex two- or three-dimensional advective/diffusive transport calculations.

Chemistry

The final major characteristic of the near-field is the water chemistry. The chemistry of the far-field groundwater, together with the hydrogeology, will define the input rate of solutes to the near-field and hence its extent of interaction with the various barrier materials and with the waste itself. A full specification of groundwater chemistry requires knowledge of at least:

(a) pH;
(b) the extent to which the water is oxidizing or reducing (*redox conditions* or Eh);

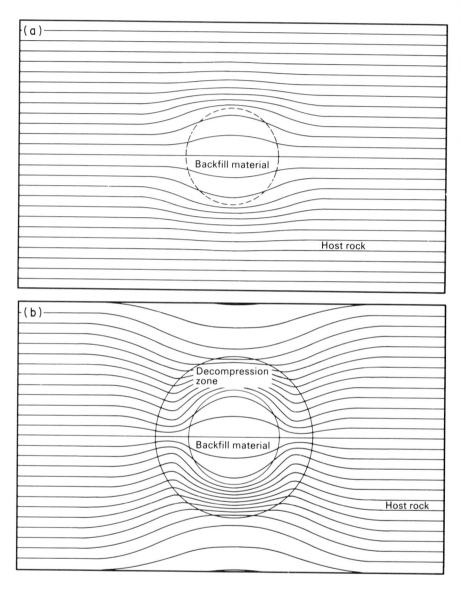

Figure 5.3(a) Theoretical groundwater flow lines, in an homogeneous isotropic medium, around a repository tunnel containing a low hydraulic conductivity backfill. (b) The same but with a more permeable decompressed zone in the rock caused by tunnelling (after Nagra, 1985). *Reproduced by permission of Nagra, Switzerland*

(c) the concentration of major ions (Na^+, K^+, Ca^{2+}, Cl^-, SO_4^{2-}, etc)

In addition further trace species might need to be defined to evaluate their reaction with particular barriers or their effect on radionuclide mobilization.

As groundwater flows or diffuses from the far-field into the near-field, its chemistry will change due to reaction with its surroundings. Even before contact with the engineered barriers, hydrothermal reaction with damaged or heated host rock may alter the water chemistry, a process associated with the formation of secondary minerals. The changes in chemistry associated with such reactions, and indeed reactions with backfill, container or waste matrix, can be evaluated by chemical thermodynamic calculations. A large number of computer codes exist which can predict the result of reaction of water of known initial chemistry with a range of solid phases at any given temperature (e.g. MINEQL, EQ3/EQ6, PHREEQE, WATEQ, etc.; see Jenne, 1981). If the timescales of the reactions are such that chemical equilibrium can be assumed, the main limitation to the use of such codes is the availability of appropriate thermodynamic data. For such calculations, basic thermodynamic data (free energies and entropies) are required for *all* reactions which may occur in the system. Even if only the most likely reactions are considered, extensive gaps in the thermodynamic database currently exist. This is especially problematic for some of the rather uncommon elements such as Se, Tc, Sn, Pd and the actinides. For such species, data either have very large associated uncertainties or are completely missing.

Apart from bulk water chemistry, trace element solubility and speciation (i.e. the chemical form of the dissolved material) can be derived from such models. The solubility limit is obviously an important constraint on the release rate of radionuclides, while their speciation (as anions, cations, or uncharged molecules) greatly alters their retardation during transport. Rates of formation or reaction of solid phases can be derived by coupling such chemical models to models of solute transport.

The chemical interaction of groundwater with various barriers and the application of thermodynamic models to predict speciation and solubility limits are discussed further in the following sections. Before this, however, some factors which may possibly affect the applicability of simple chemical models are briefly considered.

Colloids and metastable phases

Colloids can be defined in this context as particles in the nanometer to micrometer size range which can form stable suspensions in a liquid phase. Metastable solid phases are unstable thermodynamically but exist due to the very slow kinetics of their alteration into more stable products. Neither of these can be adequately taken into account in present water chemistry models. Colloids are especially problematic, as they could be mobile in the aqueous phase and thus cause barrier degradation and radionuclide transport at rates greater than expected on the basis of solubility/speciation constraints.

Micro-organisms

Micro-organisms are well established as causes of extensive material degradation, even at considerable depths (West *et al.*, 1985). Even the rather harsh environment of the near-field of a HLW repository cannot be guaranteed to be, or to remain, sterile. The effects of micro-organisms may either be direct—enhancing degradation of structural materials by enzymatic catalysis or increasing transport rates by acting as living colloids—or indirect—perturbing the near-field with metabolic by-products such as gases or organic complexing agents (McKinley *et al.*, 1985).

Organic complexants

A wide range of organic molecules may be naturally present in deep groundwaters, or may be produced in the near-field by microbial action. Such organics may more readily form stable complexes with key radionuclides than do many of the important inorganic complexants present in groundwaters. The very wide range of chemistry of such organics, together with the almost complete absence of any relevant thermodynamic data, makes the modelling of the behaviour of such species almost intractable at present.

Given this background information on the physico-chemical environment of the near-field, we can now examine the role and performance of the individual engineered barriers.

PERFORMANCE OF THE ENGINEERED BARRIERS

Backfill (or Buffer)

The backfill may fulfil a number of roles, including minimization of groundwater access to the waste package, alteration of groundwater chemistry, and retardation of solute transport. It may display plastic properties—filling any gaps formed during or after emplacement and may contribute to the even spread of near-field mechanical stresses. It also acts as an important conductor of radiogenic heat.

In fractured crystalline host rocks the most widely advocated backfill material is bentonite clay (which consists principally of sodium montmorillonite) compacted to a high density prior to emplacement, or a compacted mixture of bentonite and quartz sand which conducts heat more readily (Pusch, 1979). Bentonite displays the ability to swell as it takes up water from the surrounding rock, and in doing so would squeeze into fissures around the borehole wall and fill any other cavities, while exerting a considerable swelling pressure on the waste package itself. Since it has an extremely low permeability to water (in the order of 10^{-13} m.s^{-1} hydraulic conductivity) it effectively prevents advective flow of mobile groundwaters to the package once it has taken up water and expanded. Diffusion of dissolved material through the extensive connected pore space in bentonite occurs, but may be greatly limited by sorption processes or the small size and surface charge of the pores. At temperatures below about 100°C, bentonite tends to buffer pH in the mildly alkaline range and the high surface area of the material ensures saturation of dissolved silica (SiO_2). Ion exchange of cations in solution for sodium in the clay structure would also occur which may have further secondary effects. For example, sodium exchange with calcium in solution encourages calcite ($CaCO_3$) dissolution. This increases the concentration of carbonate in solution and may increase pH, which could alter near-field chemistry significantly. Bentonite may also directly affect the rate of canister corrosion—for example it may react with iron or steel to form layers or iron silicates which protect the canister from further corrosion.

Apart from bentonite, several other buffer materials have been proposed, such as cement grouts or concretes which also display very low permeabilities but buffer pH to much higher values (more alkaline) than bentonite and have a rigidity and ease of emplacement that may be useful in many cases, particularly for ILW. For HLW disposal in basalt it has been proposed that finely powdered basalt would be an ideal buffer (Wood and Coons, 1982) since the glassy matrix of the rock is highly reactive to the warm groundwaters which would initially be present in the near-field, and would rapidly crystallize as a very fine grained and impermeable cement-like hydrothermal reaction product, which might then seal the waste package from further water ingress. In a dry formation such as salt, the requirements for backfill performance are much less extensive than for wet conditions, being mainly concerned with ensuring stress and heat transfer. Crushed salt is probably sufficient in itself. In very dry climates where a deep repository is above the water table (e.g. in tuffs at the Nevada Test Site), it has been proposed that an air gap may be the best 'backfill', as the capillary suction in the pores of the unsaturated rock ensures that the gap will not fill up with water (Oversby and McCright, 1985).

Finally, very complex admixtures have been proposed (e.g. Beall and Allard, 1982) to tailor a buffer to act as a 'getter' which would scavenge certain leached radionuclides from solution and fix them. Such buffers aim to control the oxygen activity of passing groundwaters so that radionuclides, principally the actinides, are reduced to valence states where they are most strongly sorbed on the buffer matrix. Apart from redox control components such as metals and sulphides, and sorbing media such as clays or zeolites, these backfills might also contain minerals which are known to react with specific elements and remove them from solution. These processes and their effects are considered in detail in the modelling section. At present the stability, longevity and reactivity of such tailored buffers in the warm hydrothermal environment of the near-field are not well understood.

Backfill material will be selected on the basis of the general considerations above, but its performance under site and design specific conditions has to be evaluated. Physical and engineering properties (e.g. thermal conductivity, strength, plasticity, swelling pressure) can be readily measured by standard techniques. Evaluation of chemical reactions and backfill stability is generally more problematic. In most cases the rock/groundwater/backfill/canister system will be thermodynamically unstable and undergo slow but continuous chemical reaction until backfill and canister are totally altered into stable products. The modelling of such reactions from first principles using rigorous thermodynamic calculations is beyond present capabilities and may be further complicated by kinetic factors at lower temperatures ($\lesssim$ 100 °C). At present, therefore, performance is mainly evaluated based on empirical laboratory experiments. In order to be applicable, such experiments must rigorously duplicate the expected environment (in terms of temperature, pressure, chemical components and so on) and even then extrapolation from laboratory to geological timescales is difficult to do with confidence.

In the case of bentonite (perhaps the best studied backfill material), all relevant physical and engineering properties have been well measured, and predictions made on the basis of such measurements have been validated by laboratory and field experiments. In granitic groundwaters, the montmorillonite clay of which bentonite consists is expected to alter into another clay (illite). This alteration requires a supply of potassium ions from the groundwater which, given the low flow rates expected, ensures backfill lifetimes of millions of years, even without consideration of further limitations caused by slow kinetics (Anderson, 1984). Although this conclusion is supported by observations of bentonite behaviour in natural geological systems, alternative alteration mechanisms have been proposed and a rigorous mechanistic model for this process does not currently exist. Similarly, although simple models of bentonite/water reaction exist, they do not consider the minor components of the bentonite (i.e. other than montmorillonite), do not take kinetics into account, and are based on rather poorly specified laboratory data.

Radionuclide sorption onto bentonite has been extensively studied but in most cases experimental conditions were not relevant to those expected in repositories and the retardation predicted often agrees poorly with that observed in *in situ* migration experiments.

Container

The function of the canister (and overpack if present) in most disposal concepts is to protect the waste from groundwater for a minimum period of time during which a certain proportion of the contained radioactivity will have decayed away and any thermal transient caused by radiogenic heat will have passed. The container may also be designed for ease of handling during emplacement operations or for optimization of packing geometry in the repository. The radiation shielding effect of the container limits possible exposure to operators and radiation damage to external engineered barriers after emplacement (including the radiolysis of groundwaters).

The mechanical performance required to ensure that it will withstand emplacement operations and subsequent pressure re-equilibration generally leads to the choice of a metal for the container. More sophisticated designs (for example an inner metal canister surrounded by a ceramic overpack) are also being studied, but analysis of their behaviour is complex.

Two basic conceptual choices of container design exist (Marsh, 1982). The first aims at *corrosion resistance*, using materials which are thermodynamically stable in the repository environment, or form passivating layers of corrosion products. For example, native copper is found in some deep crystalline rock environments in Sweden, which leads to expectation that a copper canister would be stable over a period of millions of years. Although titanium would corrode quickly in this environment, it is known to form an oxide layer which limits further corrosion, which again provides a very long canister life. As very little (or no) corrosion is expected, resistant containers can be relatively thin and thus may be constructed out of expensive materials. The alternative concept is *corrosion allowance* in which container corrosion is expected, but is allowed for by designing the container to be thick enough to give the desired lifetime. In this case the container may be very massive, and made out of a cheap material such as cast iron or steel.

Much of the research on containers is involved with the practicalities of their construction and in particular, their sealing. Lid seals, welds and seams are all potential sources of weakness and likely sites for enhanced, localized corrosion. For most of the materials being considered seriously, techniques currently exist which seem capable of solving such problems. In the case of copper, for example, techniques such as electron beam welding and hot isostatic pressing have been demonstrated to provide reliable quality control.

Nevertheless, most safety assessment models incorporate terms to allow for containers which have been inadequately sealed and escaped quality control checks prior to emplacement. Such analyses can be made probabilistically, and essentially give a proportion of early container failures in the near-field source-term.

After choice of container material, it is necessary to demonstrate that desired lifetimes can be obtained in disposal systems where this is considered an important factor. In cases where the groundwater supply rate to the near-field is very low, this in itself may constrain minimum container lifetime. Otherwise the rate of corrosion under expected chemical conditions must be extrapolated from laboratory measurements. Again the total chemical system is usually too complex to be rigorously modelled at a thermodynamic or mechanistically based kinetic level.

Corrosion rate measurements may be complex in practice. For some materials, the rates of corrosion are so slow that simple weight loss measurements are impracticable on a laboratory time-scale and sophisticated indirect electrochemical techniques are required. In addition, some form of surface analysis or inspection is required to determine the spatial variability in corrosion rate. In general, enhanced local corrosion (pitting) is less important for thick corrosion allowance canisters than thin corrosion resistant materials as the pitting factor (the ratio of the maximum to the average corrosion depth) is usually observed to decrease with increasing amount of corrosion. Nevertheless, particular conditions which enhance pitting (e.g. electrical cells set up at the junction between different materials) should be carefully avoided. Measurements must be performed under realistic conditions of temperature and pressure. Apart from absolute pressure, the stress field expected on the canister may be important as corrosion may be enhanced at positions of greatest stress, and mechanical failure may occur earlier than expected due to so-called stress/corrosion cracking. Finally, the measurements should take into account the effects of the backfill (both on water chemistry and direct reaction) and, where relevant, the radiation field. For HLW in particular, the latter has two main effects—direct damage to the container structure and radiolysis of water. The 'ageing' effect on the metal caused by the low radiation field acting over long periods of time can be simulated to some extent in the laboratory by short exposure to much higher radiation fields in a reactor. For most of the materials considered, such work shows that this is not likely to be a significant problem. Breakdown of surrounding water by radiation (radiolysis) can give rise to a mixture of oxidants (such as hydrogen peroxide, H_2O_2), and reductants such as hydrogen gas (H_2). If the latter is lost, by rapid diffusion, then a net oxidising environment is produced, which generally results in enhanced container corrosion. Interaction of radiation with components dissolved in the water may result in the formation of acids (e.g. nitric acid from trapped air dissolved in the water) which again are generally corrosive. The effects of radiolysis can be quite significant for the thin corrosion resistant canisters as external radiation fields are higher and even small amounts of localised corrosion could cause premature failure. For this reason, some corrosion resistant overpacks include a radiation shield, such as lead, inside the resistant metal.

The final model for calculating canister lifetime may simply extrapolate from measured corrosion rates or also take into account the rate of supply of corrodants and radiolytic effects. A further factor which could be significant in some cases is microbial corrosion but scoping calculations indicate that its effect is small and can generally be ignored or assumed to be negligible (McKinley, *et al*, 1985b).

Apart from container lifetime, the chemical effects of its corrosion on other near-field processes may have to be considered. For example, corrosion of metal under anoxic conditions often results in the formation of hydrogen gas. For corrosion-allowance containers within a low permeability backfill such as compacted bentonite it has been calculated that this could build up extremely high pressures which could cause premature canister failure, disrupt the integrity of the backfill or even damage the surrounding rock. At present it is expected that such pressure build up will be self-limiting by decreasing the corrosion rate or causing gas flow (rather than diffusion) through the backfill, but the inclusion of a fine sand layer around the canister to decrease corrosion rates further as gas builds up is also considered (Neretnieks, 1985).

Even after water has penetrated the container it may continue to perform some barrier role. For the chemically-unreactive containers this may simply be as a physical constraint on solute transport resulting from remnant overpack and corrosion products. More reactive materials may, after failure, also act as chemical buffers and radionuclide sorbants. In particular iron oxides resulting from corrosion may buffer both pH and redox conditions (ensuring a chemically reducing environment) and strongly sorb many radionuclides (McKinley, 1985a). Quantitative evaluation of these chemical effects indicate that they could be very important, but the requisite laboratory background data are somewhat limited at present.

Waste Matrix

The main function of the waste matrix is to limit the rate of release of radionuclides after failure of the container. The various matrix materials used for conditioning HLW/ILW for deep disposal were discussed in Chapter 3. In this section we will consider the behaviour of two HLW matrices (borosilicate glass and the unreprocessed uranium dioxide spent fuel itself) and the most commonly advocated ILW matrix, cement.

For HLW, only two matrices are currently 'available'. The first is unreprocessed spent fuel, and comprises the used, irradiated fuel rods, which are inserted in the disposal canister which may subsequently be evacuated or filled with gas or lead. The second is borosilicate glass, which incorporates the waste radionuclides intimately dispersed throughout its structure. It can be formed either by in-can melting or by pouring molten glass into a steel fabrication container. It is generally assumed that during cooling and subsequent handling and transport, the manufactured blocks of glass would fracture, and hence present relatively large surface areas to any water which penetrates the engineered barriers. This is a conservative assumption used in the design of leaching experiments and in calculating release rates. These two waste types are treated separately below.

Leach testing:

Readers may come across many references to 'leach-testing' of all types of solid radioactive waste destined for geological disposal and it is worth pointing out that there are basically two types of experiment which should not be confused. The first is essentially a standard sorting technique, used to compare the overall quality of waste forms, for example batches of a vitrified waste with slightly different compositions. These are tests only, and give information on the bulk 'leachability' of a product. Various standard techniques are used, the IAEA recommended procedure being commonest. Some *dynamic* tests use continuously replenished fresh water (e.g. Soxhlet tests) as the leaching agent.

The second type of leach testing is an experimental method which attempts to replicate realistic disposal conditions. As discussed later, the disposal environment will be characterized by virtually zero groundwater flow, so these experiments are generally closed-system, static leaching tests. Data are produced in the form of individual element concentrations in solution as a function of time, temperature, solid to fluid ratio, and so on. This second type of experiment is the only reliable means of providing data for release modelling and, as many authors have pointed out (e.g. Ogard and Bryant, 1982; Savage and Chapman, 1982), data on bulk leach rates from flow-through tests should not be applied to realistic safety assessments. The leaching of various waste forms has been very intensively studied and is now quite well understood. This can be illustrated by considering borosilicate glass in some detail.

Borosilicate glass behaviour:

While a variety of glass formulations have been studied as containment media, including phosphate and syenite glasses, borosilicate compositions have demonstrated greater stability or flexibility. Their formation temperatures are quite low, they are able to incorporate the bulk of the HLW components satisfactorily, and their leach behaviour is well understood. A typical waste glass composition is shown in Table 5.1. The major glass network forming complex is the SiO_2 molecule, with silica comprising up to about 50 wt per cent of the glass and dominating the dissolution behaviour of the glass in water.

Table 5.1   COGEMA specification (1982) for the composition of high-level borosilicate waste glass

Component	Weight %	Actinides	g/Container
SiO_2	45.2	Am	423
B_2O_3	13.9	Cm	33
Al_2O_3	4.9	Pu	80
Na_2O	9.8	Np	573
CaO	4.0	U	1920
Fe_2O_3	2.9		
NiO	0.4		
Cr_2O_3	0.5		
P_2O_5	0.3		
ZrO_2	1.0		
Li_2O	2.0		
ZnO	2.5		
FP oxides[a]	11.1		
Actinide oxides	0.9		
Metallic particles	0.7		

Container weight $\sim$480kg; glass weight = 405 kg/container.
[a] FP = fission product.

Borosilicate glasses react with water very slowly at low temperatures (Marples *et al*, 1980; Boult *et al*, 1978), but can be highly reactive at elevated temperatures and pressures (Savage and Chapman, 1982). The mechanism of waste dissolution changes with increasing temperature, and as a function of time. At temperatures approaching 200 °C the glass begins to break down quickly under hydrothermal conditions.

However, the breakdown of the glass matrix may not be the principal factor which controls the release rate of radionuclides into the near-field groundwaters. As the waste breaks down, a wide variety of secondary products are formed. These can be amorphous materials, colloids or stable mineral phases, depending on the temperature and the groundwater composition. The solubility of any radionuclide is thus dependent on the extent of its incorporation into these new materials and the relative stabilities of each in the groundwater environment. These in turn will vary with the rate of turnover of water, and with temperature and redox conditions. Thus radionuclide mobilization, either in solution or particulate/colloidal form, will be a rather complex function of the thermodynamics of a number of reactions in the near-field. This argument is true for any waste form where groundwater access is limited by the barrier system and holds for ILW types as well as HLW types. Tracking the behaviour of individual radionuclides up to their point of mobilization, in the manner described above, is known as the 'fate of nuclides' approach, and has been convincingly advocated by Apted and Myers (1982) and Savage and Robbins (1983).

The general approach is summarized in Fig. 5.4 (after Savage, 1984, and Apted and Myers, 1982) which illustrates the change in concentration in solution of a radionuclide released from any metastable wasteform, such as glass, as a function of time. Up to time t_C, which is very dependent on the temperature of the system, the rate of water exchange and the surface area of the waste form being leached, a wide variety of 'leach rates' can be produced; leach rate being equivalent to the gradient of the tangent to the curve at any given time, t_A, t_B, etc. These are clearly only applicable to that time, and a general trend of decreasing leach rate can be seen, as the tangents flatten out. All of these leach rates will overestimate releases. When the concentration curve flattens out the radionuclide will be present in solution at its solubility limit, with respect to whatever phase is dominating its behaviour at that time. With increasing time the whole near-field system will progressively readjust to replace metastable reaction

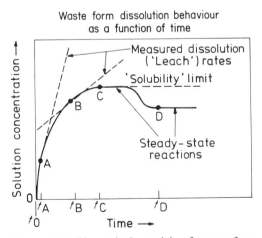

Figure 5.4 Theoretical model of waste-form dissolution as a function of time in a low-flow environment, expressed as the concentration of a particular radionuclide in solution in the groundwater as a function of time; after Savage (1984). See text for explanation

products with more stable phases, and the consequent solubility limits of individual radionuclides will decrease in a stepwise fashion (e.g. to time t_D). The initial 'solubility' limit portrayed thus represents a conservative figure for radionuclides in solution at any time after disposal. An important feature of this approach is that experiments must aim to characterize fully the solid phase chemistry and aqueous phase speciations in order to interpret and predict the overall reaction path.

Taking one step back from this approach, simple empirical calculations can be made of mobilization rates in a near-field where water availability is considerably limited by the barriers and the surrounding far-field. This was attempted, using very simple solubility criteria, by Chapman *et al*, (1982). A further example of the use of this approach is presented in the detailed near-field model considered later. Depending on the leaching mechanism chosen, and the degree of significance attached to saturation effects, this highly empirical approach predicted very protracted waste mobilization episodes, suggesting that the waste glass and the secondary waste products will have extremely long lifetimes, in the order of millions, or tens of millions, of years, with releases taking place at very low concentrations.

Spent-fuel behaviour:

Spent fuel is a very different waste form to borosilicate glass. First, its content of nuclear-reaction by-products is more dilute, since the fuel has not been reprocessed and these radionuclides extracted and concentrated. Second, the waste matrix is either uranium metal or oxide, rather than glass, and the dissolution behaviour of the waste is thus largely controlled by the solubility of uranium in water. The matrix is itself radioactive and, when integrated over long time periods, radiation effects such as alpha radiolysis may be more significant. Third, the thermal output of spent fuel shows different characteristics, being significant over substantially longer periods of time (up to 10 000 years), rather than exhibiting steady decay over about 1000 years. In addition, the presence of very labile elements (such as Cs and I), which diffuse through the waste matrix and concentrate on the outer surface during the operational life of a fuel element, must be accounted for when studying the leaching of the waste.

Leaching is controlled by the solubility of uranium which is very dependent on redox conditions and the concentration of complexing species (ligands) such as carbonate in solution (e.g. Allard, 1983). In reducing conditions this solubility is very low and, in addition, deep groundwaters are usually already saturated with naturally occurring uranium. In much the same way as glass dissolution described above, individual radionuclide mobilization rates are controlled by the nature of surface layers formed, and are highly dependent on groundwater pH and temperature (Wang and Katayama, 1982). Models of radionuclide mobilization often avoid the complex thermodynamic consideration of secondary reaction products by making the generally conservative assumption that releases are congruent with the matrix dissolution. An exception is immediate releases, such as the 'instant' iodine, or caesium mentioned above. These elements concentrate at grain boundaries or the inner surfaces of cladding during reactor operation.

Long-term leach tests on spent fuel in natural and synthetic groundwaters have been performed mainly in Canada and Sweden, while in the USA work has concentrated on their hydrothermal stability in brines (Gray and McVay, 1983) related to disposal in salt formations. As noted above, the UO_2 matrix dissolves more rapidly under oxidizing conditions, and these may be produced in the very near-field as a result of radiolysis of groundwaters. Leaching experiments on spent uranium dioxide fuel from CANDU reactors at temperatures of 25–30 °C have been underway for 8 years (Stroes-Gascoyne et al., 1986), using distilled water and tap water.

Matrix dissolution rates are of the order of 10^{-6} to 10^{-9} per day, with uranium concentrations in solution well below theoretical saturation limits. Crystalline reaction products are formed on the fuel surface, and the mobilisation rates of more labile species such as Cs decrease very rapidly with time, to a presumably steady-state level, reminiscent of the process outlined earlier for glass dissolution. Long term tests in Sweden (Forsyth et al, 1986) in synthetic granitic groundwaters show that Sr behaves similarly, although the more significant radionuclides (U, Pu, Cm, Ce and Eu) enter solution more slowly, reflecting solubility limitation controlled by their precipitation or adsorption on to the fuel surface, again parallel to the glass dissolution mechanism. Earlier Swedish work also demonstrated that both Sr and U were present in the aqueous phase as colloidal material; up to 50 per cent of the amount in the leachate in the case of Sr.

Cement waste-form behaviour:

Owing to the heterogeneous nature of the solid ILW fragments incorporated into the matrix material, and the variety of matrices in use, it is not possible to give a simple picture of ILW leach behaviour. Here, we consider only cement matrices in detail. For ease of modelling it is often pessimistically assumed that the waste radionuclides are homogeneously distributed throughout the porous cement matrix; that is, the metallic, plastic or resin fragments offer no additional barrier to leaching. In these circumstances radionuclide releases are controlled by individual elemental solubility limits and diffusion rates in the matrix pore water. The very high pH of the cement pore-water (11.5 or greater) greatly limits the solubility of many elements.

Chapman and Flowers (1986) compiled solubility data of all the relevant radionuclides in ILW under the high pH conditions relevant to cement waste forms. The solubilities of the significant long-lived radionuclides are very low, with the exception of ^{226}Ra, ^{14}C, ^{135}Cs and ^{129}I. Experimental studies on the leaching of various typical ILW types at temperatures up to 90 °C, and with or without protection from atmospheric carbon dioxide, are summarized by Amarantos et al. (1985). A high mobilization rate of Cs in ordinary Portland cement (OPC) matrices was observed, with diffusion out through the pore structure of the cement being one of the controlling mechanisms. As in the case of glass, however, the production of secondary phases was seen to have a marked effect on the mobilization of some elements; in this case in partially blocking the pore structure. For example Sr can co-precipitate with Ca. Various silica, slag, and pulverized fuel ash (PFA) additives produce similar effects for Cs, and solubility once again is seen to be controlled by interaction with secondary products as well as matrix materials. In the case of cement waste forms, it appears that a much wider variety of metastable phases can influence radionuclide behaviour.

Since the maintenance of a high pH regime is important to ensure low solubilities, Atkinson *et al.* (1985, 1986) studied the likely long-term evolution of cement pore-water pH, and the general longevity of cements in disposal environments. The pH was predicted to fall in a stepwise fashion (Fig. 5.5), but would not fall below 10.5 for at least a million years. The engineering lifetime of the concrete is likely to be considerably less, owing to sulphate attack, Ca leaching or reactions with the aggregate material. There seems to be little doubt, however, that for a cement waste form with concrete containers and engineered

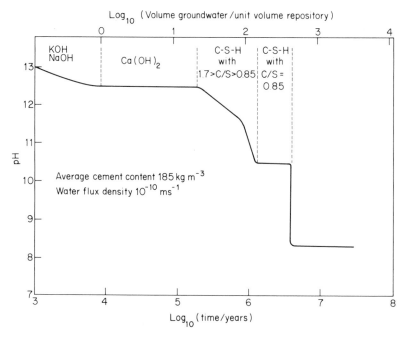

Figure 5.5 The estimated time dependence of cement pore-water pH, for a model spherical repository of 20 m radius containing 185 kg/m³ Portland cement, situated in a low groundwater flow environment (after Atkinson, 1985). At times up to 10^4 years the chemistry is dominated by sodium and potassium hydroxides in the cement, by calcium hydroxide up to 10^5 years, and by calcium silicate hydrates thereafter, until the pore system has been effectively 'flushed'. A high pH is maintained for at least 10^6 years

barriers, near-field pH will remain high even if some structural integrity is lost after a few thousand years.

Other effects on leaching

Although not considered in detail above, high radiation dose rates have little direct influence on the leaching of most HLW matrices (although radiolysis of groundwater or backfill pore-waters may have similar effects to those considered in the section on container behaviour). However, very significant effects can arise from the much lower dose rates to some ILW matrices. For example, radiolysis of bitumen can produce a range of gaseous products (e.g. hydrogen, methane) which could not only burst the canister and overpack, but also open up a larger system of internal pores, which would increase subsequent waste leaching.

Geological and archaeological analogues provide good evidence for the stability of most of the matrices considered for both HLW and ILW (see Chapter 11) but for many of the latter there are also indications of their susceptibility to microbial degradation (West *et al.*, 1985). As yet, very little quantitative data on such biodegradation exists but work in this area is receiving increasing attention.

This section has examined the processes involved in the near-field and the methods used for their quantification. Next we consider how to put all this information together to calculate the resulting releases into the far-field.

MODELLING OF RADIONUCLIDE RELEASE FROM THE NEAR-FIELD

Given the background provided in the previous section, we can now follow through the calculations used in repository safety assessment to evaluate releases from the near-field. These are used as input for the far-field migration calculations considered in the next chapter. First of all we look in detail at one specific near-field layout which was recently proposed for a Swiss deep HLW repository. The general principles can be applied to any deep repository near-field for all types of long-lived wastes, but some specific issues emerge which are briefly examined. Finally the problems and limitations associated with present models are discussed.

The Swiss near-field model

The base case of the recent Swiss analysis (NAGRA, 1985) envisages disposal of vitrified HLW in a cast steel overpack. The containers are emplaced horizontally into tunnels which are backfilled with compacted bentonite. The repository is at a depth of $\sim 1200\,\mathrm{m}$ in granite. Deep groundwater flow is heterogeneous, occurring mainly in widely separated, subvertical disturbed zones in the rock. The calculated water flux through the entire repository is 4200 l each year. This water is slightly reducing chemically, and at near neutral pH. The ambient temperature at repository depth is $\sim 55\,°C$ and the radiogenic thermal transient

88

gives a maximum temperature of ~ 160 °C inside the backfill and ~ 100 °C at the disposal tunnel wall. These high temperatures drop off within the first few decades, and are insignificant after about 500 years as far as near-field analysis is concerned. The near-field around a single container is illustrated in Fig. 5.6 and the quantities of materials involved listed in Table 5.2.

Calculations of the type described in the previous section (McKinley, 1985b) indicate that:

(1) The bentonite will be mineralogically stable for $> 10^6$ years and its chemical and mechanical properties will be retained for such a period
(2) After corrosion of a 5cm thick 'allowance', the container will fail

Table 5.2 Material inventory (per waste container) in the near-field of a reference Swiss high-level waste repository (from McKinley, 1985b)

Material	Volume (m³)	Mass (kg)
Glass	0.15	405
Steel-fabrication container	0.01	75
Fabrication void	0.03	—
Canister	0.9	6.5×10^3
Backfill	52.8	—
(a) Bentonite (dry)	32.7	8.8×10^4
(b) Pore space (water-filled)	20.1	2.0×10^4

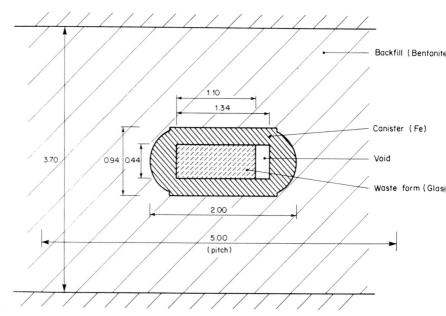

Figure 5.6 Geometry of HLW container emplacement (dimensions in metres) for the Swiss in-tunnel deep repository concept (after Nagra, 1985). *Reproduced by permission of Nagra, Switzerland*

mechanically. This corrosion will take at least 1300 year (rounded down to 10^3year in the safety analysis). The low solubility of the corrosion products ensures that they exert a chemical buffering role for $\gtrsim 10^6$ years.

(3) At the time of canister failure, the glass will be fractured to an extent which increases its area by a factor of approximately 12 times its original value. The rate at which the glass breaks down is 10^{-7} g/cm^2 of surface area/day

(4) Chemical buffering reactions in the bentonite and canister ensure that the pH of pore water is slightly alkaline (in the range 7–8.5) and chemically reducing. The solubilities of particular elements of interest under these conditions are listed in Table 5.3.

Combining all this information, we can build our first simple model of release. Table 5.4 lists the main fission/activation product radionuclides in the HLW,

Table 5.3 Solubilities of some key elements in the near-field conditions of the reference Swiss high-level waste repository (from McKinley, 1985b)

Element	Log. solubility (mol/1)
(a) Literature data	
Be	-4
Cs	high
Ca	-2
C	high
Ho	-8
I	high
Pb	-6
Mo	-2
Ni	-4
Nb	-8
Pd	-8
Ra	-8
Rn	high
Sm	-8
Se	-8
Ag	-4
Sr	-4
Sn	-9
Zr	-9
(b) Extrapolated from MINEQL/EIR-Data	
Tc	-7.5
Ac	-2
Th	-5.5
Pa	-5.5
U	-9
Np	-8
Pu	-6.5
Am	-2
Cm	-2

their half-lives, and their inventory in each canister. As the canister lasts 10^3years, we can take decay during this period into account and the table shows that within this time the quantities of some short-lived nuclides have already become insignificant. Over the long periods involved, it is considered that the waste dissolves congruently (see previous section) which, immediately after canister failure, would release in the order of 10^{-5} of the inventory each year. If the bentonite/canister barriers are ignored at present and all the radionuclides released are assumed to dissolve in the water flux of 0.7 l/canister/year, their concentrations in solution would be as given. As we discussed previously, however, these releases may be further constrained by limiting solubilities, and the concentration which would saturate the reference water flux is shown in the final column (where more than one isotope of the same element is present, total solubility is divided between them in proportion to their total inventories). It can be seen that the concentration which would actually be released to the far-field is the lower of the values in the last two columns and this is shown underlined. Releases of ^{137}Cs, ^{90}Sr, ^{108m}Ag, ^{121m}Sn, and ^{151}Sm are all negligible.

Table 5.4   Fission activation product inventory 1000 years after disposal in the Swiss reference HLW repository. Comparison of constraints on release set by the matrix dissolution rate and elemental solubility limits (from McKinley, 1985b)

Nuclide	Half-life (y)	Inventory (mol)	Release rate limited by congruent dissolution		Solubility limited release rate	
			(mol/y)	(Bq/y)	(mol/y)	(Bq/y)
^{10}Be	1.6E6	2.6E-5	*5.1E-10*	*4.2E0*	7.1E-5	5.9E5
^{14}C	5.7E3	1.9E-5	*3.7E-10*	*8.5E2*	high	—
^{41}Ca	1.3E5	8.7E-5	*1.7E-9*	*1.7E2*	7.1E-3	7.2E8
^{59}Ni	8.0E4	1.1E-2	*2.2E-7*	*3.6E4*	7.1E-5	1.2E7
^{63}Ni	1.0E2	1.8E-6	*3.5E-11*	*4.6E3*	1.2E-8	1.6E6
^{79}Se	6.5E4	9.3E-2	1.8E-6	3.7E5	*7.1E-9*	*1.4E3*
^{90}Sr	29	6.1E-10	*1.2E-14*	*5.5E0*	7.1E-5	3.2E10
^{93}Zr	1.5E6	1.1E1	2.2E-4	1.9E6	*7.1E-10*	*6.3E0*
^{94}Nb	2.0E4	3.9E-4	7.6E-9	5.0E3	*7.1E-9*	*4.7E3*
^{93}Mo	3.5E3	6.6E-6	*1.3E-10*	*4.9E2*	7.1E-3	2.7E10
^{99}Tc	2.1E5	1.1E1	2.2E-4	1.4E7	*2.3E-8*	*1.4E3*
^{107}Pd	6.5E6	2.5	4.9-5	1.0E5	*7.1E-9*	*1.4E1*
^{108m}Ag	1.3E2	2.8E-8	*5.5E-13*	*5.6E1*	7.1E-5	7.2E9
^{121m}Sn	50	4.2E-11	8.2E-16	2.2E-1	*8.3E-20*	*2.2E-5*
^{126}Sn	1.0E5	3.6E-1	7.0E-6	9.3E5	*7.1E-10*	*9.4E1*
^{129}I	1.6E7	1.8E-3	*3.5E-8*	*2.9E1*	high	—
^{135}Cs	2.3E6	3.2	*6.2E-5*	*3.6E5*	high	—
^{137}Cs	30	2.0E-8	*3.9E-13*	*1.7E2*	high	—
^{147}Sm	1.1E11	1.5	2.9E-5	3.5E0	*7.1E-9*	*8.5E-4*
^{151}Sm	93	7.0E-5	1.4E-9	2.0E5	*3.3E-13*	*4.7E1*
^{166m}Ho	1.2E3	5.4E-6	*1.1E-10*	*1.2E3*	7.1E-9	7.8E4

Although extremely simplistic, this type of approach is widely used in calculating release functions. Computer models can readily calculate releases of all nuclides as a function of time taking into account changes in glass surface area during leaching, and complex decay chains. A typical example is shown in Fig. 5.7. It may be noted from this figure that even though the glass matrix has been totally destroyed within 150 000 years, the limited solubility of the radionuclides from the secondary reaction products in which they are by this time incorporated, may greatly extend release times. The model above is very conservative as it ignores the diffusive resistance of the bentonite. We could improve it by assuming radionuclides are released into a small volume of water inside the canister (corresponding to the original void space, say) and that diffusive transport to the tunnel wall then occurs. If we assume that a water volume of 0.7 l is distributed around the wall (as a 'mixing tank'), and this is changed each year, this could be considered a reasonably conservative boundary condition. For the diffusion calculation, interaction between radionuclides and the bentonite is modelled very simply by a partition coefficient which assumes a constant ratio of concentrations in the solid and liquid phases, i.e.:

$$K_d = C_R/C_w$$

where K_d is the partition coefficient or distribution coefficient (e.g. m^3/kg);
  C_R is the radionuclide concentration on the rock (e.g. Bq/kg);
  C_w is the radionuclide concentration in the pore water (e.g. Bq/m^3).

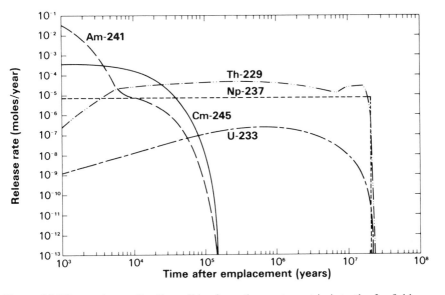

Figure 5.7 Direct release of radionuclides from the waste matrix into the far-field as a function of time after waste emplacement in the repository for the reference case water flux of 4200 l/year through the entire repository (after Nagra, 1985; see text for explanation). *Reproduced by permission of Nagra, Switzerland*

This distribution coefficient can be transformed into a retardation factor (R) by taking into account the porosity and density of the rock. The R value corresponds to the ratio of the speed of transport of an ideal, unretarded ion to that of the radionuclide concerned. This concept is of major importance in the far-field and is discussed in more detail in the next chapter. K_d and R values for some relevant radionuclides are given in Table 5.5. The final factor which has to be taken into account is radioactive decay during transport.

The maximum release rates for a number of fission/activation products calculated by a simple one-dimensional diffusion model are shown in Fig. 5.8 as a function of the 'nuclide characteristic' product of decay constant and retardation factor (λR). Given natural background concentrations of radionuclides (Chapter 1), releases below about 10 Bq/l can probably be considered negligible. Hence, out of the very large number of fission/activation products in the original waste, only [135]Cs, ^{75}Se, [107]Pd, ^{99}Tc, and [126]Sn are mobilized into the far-field in significant quantities. The time profiles of release for actinide decay chains can be

Table 5.5   Reference near-field model for Swiss HLW repository: radionuclide sorption data for the bentonite backfill (after McKinley and Hadermann, 1984)

Element	Kd (m^3/kg)	R^a
Ac	5	2.3 E4
Am	5	2.3 E4
Be	0.01	4.6 E1
Cs	0.2	9.0 E2
Ca	0.2	9.0 E2
C	0.005	2.4 E1
Cm	5	2.3 E4
Ho	2.5	1.1 E4
I	0.005	2.4 E1
Pb	1	4.5 E3
Mo	0.005	2.4 E1
Np	1	4.5 E3
Ni	1	4.5 E3
Nb	2.5	1.1 E4
Pd	0.005	2.4 E1
Pu	5	2.3 E4
Pa	1	4.5 E3
Ra	0.2	9.0 E2
Rn	0	1
Sm	2.5	1.1 E4
Se	0.005	2.4 E1
Ag	0.5	2.3 E3
Sr	0.2	9.0 E2
Tc	0.25	1.1 E3
Th	1	4.5 E3
Sn	0.05	2.3 E2
U	1	4.5 E3
Zr	5	2.3 E4

^{a} $R = 1 + \frac{1-\varepsilon}{\varepsilon} \rho Kd$; ε = porosity (0.38); ρ = specific density (2.76×10^3 kg/m^3)

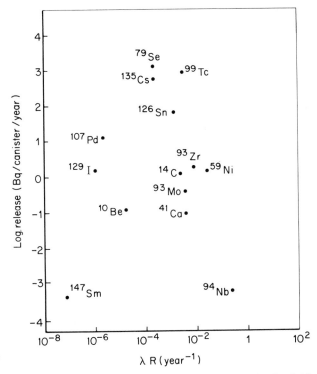

Figure 5.8 Rates of release of radionuclides to the far-field calculated for the Swiss HLW disposal concept, using a simple diffusive transport model (after McKinley, 1985b; see text for explanation). *Reproduced by permission of Nagra, Switzerland*

calculated similarly and that in Fig. 5.9 can be compared with the preceding (Fig. 5.7).

Release models for other near-field designs

The numerical values in the previous section are obviously 'concept-specific' but the general approach can be applied to any near-field design. The essential parameters required are the functional lifetimes of individual barriers, release rate from the matrix, radionuclide solubility limits in regions of different chemistry, and retardation rates during migration within the near-field. Some factors not considered in this basic model which might, however, be important for other layouts are now considered.

For concepts in which the engineered barriers do not act as chemical buffers, radiolysis may be very important. As shown previously, radiolysis is particularly significant in the case of spent fuel, and Swedish calculations indicate that for disposal in a copper canister oxidizing conditions caused by radiolysis may penetrate up to about 50 m into the surrounding rock (KBS, 1983). For the far-field migration model used, this was a significant proportion of the groundwater

94

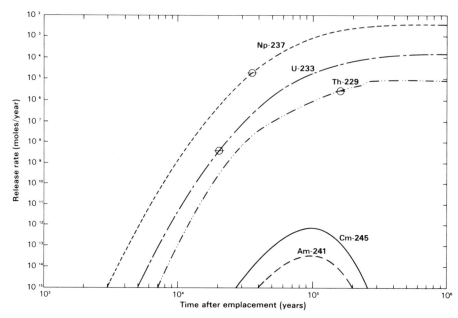

Figure 5.9 Diffusion controlled actinide release rates into the far-field as a function of time after disposal, for the whole repository (Swiss HLW concept), from Nagra, 1985. The points indicate where maximum removal is limited by the solubility of the radionuclide in 4200 l of incident groundwater from the far-field (see text for explanation). *Reproduced by permission of Nagra, Switzerland*

flow path. The evolution of this oxidizing region with time is important, as the solubility of the waste matrix and many important radionuclides is higher under oxidizing conditions, and the extent of radionuclide retardation is less. In addition, the decrease in solubility of some radionuclides at the boundary between oxidizing and reducing conditions could give rise to colloid formation—a problem which is considered in the next section. Heterogeneous distribution of radionuclides can also require variations from the basic model. The 'prompt' release of labile elements such as I and Cs from spent fuel due to their accumulation at the waste surface was mentioned previously and has to be taken into account in calculations. For ILW the chemical form of waste elements within the immobilization matrix may be very heterogeneous, even within individual wastes such as ashes, concentrates, plastics, ion-exchange resins, contaminated structural materials and so on. At present insufficient data exist to model radionuclide release rigorously and hence, especially for porous matrices such as cement, no matrix breakdown limit is assumed and releases are constrained only by solubility limits and internal diffusion rates.

Finally, temporal changes in the properties of individual barriers may have to be taken into account. For example, the initially very high pore water pH in cements will decrease with time as interaction with groundwater occurs, and the matrix permeability also changes as its physical structure degrades. Although these processes occur gradually, models generally simplify this to a stepwise progression in which the simple calculation chain is repeatedly applied for different time periods during which all properties are taken to be constant.

LIMITATIONS AND PROBLEMS

The models considered above are all obviously great simplifications of reality and the limitations of their application must always be kept in mind. For example, all analyses available assume that releases from a single canister can be treated in isolation and multiplied up to give total repository releases. An alternative approach involves associating distribution of properties—e.g. variations in canister failure time which can average out releases over a long time period (e.g. KBS, 1983; Apted et al, 1985). Processes could also be envisaged in which releases from one canister could affect others. In the case of a distribution of canister failure times, radiolytic oxidants might cause accelerated failure of containers downstream of previously failed ones in a kind of domino effect. Such interactions could also be favourable when one considers processes which are limited by solute supply from groundwater or solubility-limited releases.

Another tacit assumption in the model chain considered is that all transport occurs in solution and is thus controlled by solubility limits. If significant transport of colloids or suspended particulates occured, the results of the model would be completely invalid. Formation of colloids can result from breakdown or erosion of barriers, radiolysis reactions or precipitation at the boundary between zones of different chemistry (e.g. at radiolytic redox fronts). Colloids may also be supplied naturally by the groundwater. Few data are available, however, on the likely concentration, mobility or stability of such colloids and they are often disregarded on this basis—assuming any effects involved are masked by other large conservatisms in the analysis.

The only information available on colloid migration in the far-field applies to aquifers with large open pores, where transport of colloids, suspended particles and micro-organisms is well established. Such materials can even be used as groundwater 'tracers'. In more relevant formations few, if any, data are available. Although colloidal size particles (i.e. those less than 1 micron in size) can be detected and measured in water samples from deep boreholes, there are strong suggestions that these may be artefacts of the sampling process. Some tracer migration experiments have shown mobility of quite large organic molecules in crystalline and argillaceous rocks, but significant long-range colloid transport has yet to be demonstrated.

The pore structure of the engineered barriers may act like a molecular filter, preventing movement of colloids. Compacted bentonite, in particular, is often assumed to prevent colloid migration completely and, while this has not yet been experimentally proven, their mobility seems to be at or below the present limits of measurement (e.g. Torstenfelt, et al., 1982). In this case, if all expected formation mechanisms occur within this barrier, near-field radionuclide releases are unaffected. They may even be decreased if the immobilized colloids contain significant quantities of key radionuclides.

Apart from problems caused by colloids, the use of solubility limits also assumes that these values can be measured or predicted by using chemical thermodynamic models. For many radionuclides relevant data are extremely sparse and some values used are little better than educated guesses. The

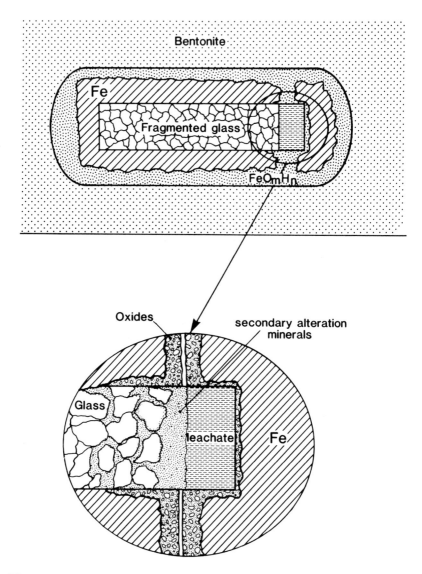

Figure 5.10 Schematic diagram of the near-field in the Swiss HLW repository, after canister failure has occurred, assuming realistic evolution as opposed to the conservative models discussed in the text (after McKinley, 1985a). *Reproduced by permission of Nagra, Switzerland*

applicability of current thermodynamic models to low temperature groundwaters is also debatable at present (Lindberg and Runnells, 1984). Finally, perturbations caused by micro-organisms have not been considered. Apart from biodegradation of engineered barriers, such organisms could also produce metabolic by-products which considerably enhance the solubility and mobility of some radionuclides. Relevant background data are very limited, but simple calculations indicate that such effects would not be significant for the Swiss HLW concept previously described (McKinley et al., 1985). Shallower repositories and ILW types are more likely to be perturbed, in particular those containing large amounts of organic wastes or bitumen, but no detailed analysis has been carried out so far.

Given the limitations and problems above, it is important to evaluate the extent of 'conservatism' in near-field models. We also need to evaluate the relative performance of individual barriers by means of sensitivity analysis, and to compare different disposal concepts. This requires the near-field models to be as realistic as possible, even though their development is still at an early stage. For illustrative purposes, however, we can again consider the Swiss HLW case. Even the most realistic model yet used, which incorporates diffusion in the backfill, has failed to consider in detail exactly what the physical system would look like (Fig. 5.10). Glass degradation would form extensive suites of secondary minerals which incorporate radionuclides and form a diffusion barrier. Any cracks through the canister which occurred during its mechanical failure would be infilled by corrosion products or swelling bentonite and form low permeability 'bottlenecks' to radionuclide transport. Even after corrosion of all the canister material, the corrosion products would probably have very low permeability and a high retardation capacity for many radionuclides. In the Swiss case most of the water flows in widely separated disturbed zones in the far-field and may not be uniformly available throughout the near-field. Hence the three-dimensional diffusion path through the backfill may be much larger than that in the simple 'one-dimensional' calculation considered previously. The total water flux through the repository occurs mainly between disposal tunnels and hence the volume available for saturation may be very low and further limited by swelling of bentonite into any open porosity of the tunnel wall.

Development of 'realistic' near-field models is now an important research area, and the limited results to date indicate that under expected conditions, applying a realistic cut-off time (say a million years), the releases from the near-field are negligible for many repository concepts. Such results very much strengthen the whole multibarrier concept.

CHAPTER 6

The Far-field: Radionuclide Migration

The 'far-field' volume of rock, the most important of the barriers to waste migration, extends from the 'near-field' disposal zone of the repository itself back to the Earth's surface. This chapter examines in more detail the function of the geological barrier in deep disposal concepts, the various rocks and geological environments which have been proposed as hosts for long-lived waste disposal, the factors which control the migration of radionuclides through these rock formations, and the assembly of a complete 'normal-case' release model for use in performance assessment. The quantification of parameters and understanding the interactions of geological phenomena in the far-field are the basic objectives of this, fundamentally geological, research area, in order that predictive performance models can be developed and tested.

FUNCTION OF THE GEOLOGICAL BARRIER

The basic requirement of a waste repository is that the combination of its natural and engineered barriers should act to control the release of radionuclides to the biosphere. The far-field acts as a massive physical and chemical buffer to processes in the near-field, principally by controlling geochemical fluxes dominated by the rates of water movement. The near and far-fields are primarily coupled by exchange of solutes (e.g. corrodants and complexing agents from the far-field, and corrosion products and leached radionuclides from the near-field). Once radionuclides are mobilized into the far-field, then the two main factors of significance in a safety assessment are:

(a) the length of time for the radionuclides to reach the biosphere, and
(b) the concentrations in which they arrive, which are very dependent on the nature of the surface water body or aquifer into which they are released.

These two 'far-field' factors are controlled by:

98

(a)   the pathlength and velocity of water-borne migration (by both advection and diffusion) through the host rock, overlying formations and superficial geology

(b)   the physical and chemical environment along this path, and in particular the ability of the rock to retard the rate of movement of individual radionuclides and dilute and disperse them in groundwaters.

Concealed within these two factors are a host of influential processes and parameters which must be understood and quantified under specific circumstances for the purposes of a migration model. In attempting to define what might be suitable sites for a repository the aim is to find an environment whose properties give a good balance or combination of these important features. In all the subsequent discussion it is assumed that leaching by and transport in groundwaters is the 'normal' mechanism by which the wastes eventually reach the biosphere. When the problem is seen in this light it follows that we must consider not just the rock alone but the total geological environment. This includes the rock formations, groundwaters and surface cover, and their behaviour as affected by climate, terrain, geological evolution and the effects of the waste itself, all of whose properties vary with time. Some of these factors are completely *site-specific* (i.e. related only to geographical position, regional geomorphology or local rock properties and structures) while others, particularly some physicochemical properties, are common to certain types of rock or certain environments, and are said to be *generic*.

The purpose of this section is to evaluate what have broadly been termed 'geological requirements' and to see how their application has led to both the choice of generically suitable rock types and geological environments and eventually, in many countries, to the definition of potentially suitable areas of land where deep disposal might take place.

The earliest comprehensive work relates to disposal of HLW. Three principal rock types (argillaceous or 'clay-rich' units; hard crystalline igneous and metamorphic rocks; and evaporites—principally halite formations) most adequately fulfil the needs of a host unit for HLW disposal, and since their identification several years ago, they have dominated international research programmes. This selection was made partly owing to their ability to accept heat-emitting wastes, but chiefly on hydrogeological grounds. It was considered that since they are all of low permeability they would have very low rates of water through-flow, in terms of both volume and velocity. The task has now become one of quantifying and testing these largely intuitive notions, often at specific sites. In particular it has become clear that the properties of the large scale groundwater regime in which the host formation lies are as important as those of the host rock itself. This is especially the case for evaporite and clay formations lying in thick sequences of mixed sedimentary rocks.

Evaporite deposits, commonly referred to as 'salt' in the literature, were thought to be completely impermeable, and also to have the advantage of being plastic enough to allow self-sealing of fractures by creep in response to heat or stress. Argillaceous rocks, although very diverse in physical properties, ranging from plastic clays through to hard, well-bedded and often fractured units, are generally characterized by low permeability and high sorption capacity (and hence radionuclide retardation potential). Massive bodies of hard, stable, crystalline rocks can have considerable vertical and lateral extent, are easy to construct underground facilities in, and are often very poorly permeable. More recently volcanic tuff deposits have been added to this group. In addition all the rock types, with some exceptions in the argillaceous and evaporite classes, are thermally stable under the heat load imposed by the wastes and can conduct and diffuse the heat away over very long periods of time.

The arrival of deep disposal concepts for other long-lived ILW types has largely been advanced on the back of the early HLW studies, and consequently very similar rock formations in similar environments are being studied for these non heat-emitting wastes. Without the added problem of finding a repository host rock with adequate thermal stability, the emphasis is leaning in some countries towards wider varieties of mixed sediments with good regional hydrogeological properties. These will be discussed later. The three groups of rock (generically known to the 'waste people' as crystalline, argillaceous and evaporite) are sufficiently broad categories to encompass almost all formations with physical properties which might prove suitable for deep disposal of long-lived wastes. Consequently they will be retained throughout the subsequent discussion. Since the mechanism of groundwater movement is different in the three generic rock groups, owing to their different hydrogeological properties, the factors to be investigated vary in each case. The following sections provide brief introductions to these rock types.

Crystalline rocks

The term 'crystalline rocks' has crept into the waste disposal nomenclature as meaning hard, massive igneous or metamorphic rocks such as granite, gabbro and basalt. The common feature of all these rock types is that they have very low total porosity, and are generally jointed and fractured (on the scale of tens of metres down to centimetres) (Fig. 6.1). Here we use the terms fissure and fracture quite loosely to apply to any physical discontinuity in the rock extending from small cracks to large shear zones. Although the contribution of the fractures to the porosity of these rocks is small, water flow through the rock is dominated and controlled by these fissures, in terms of their apertures, orientations, numbers, degree of connectivity and so on. Major fractures (faults, crush zones and other features which can be traced over many hundreds of metres and sometimes many kilometres) occur at irregular intervals, are often nearly vertical or horizontal, and can be either more or less permeable than the surrounding rock. With the exception of basalts the crystalline rocks tend to occur as massive deep-seated plutons or extensive regions of intensely metamorphosed rocks which may be buried under younger sediments, or outcrop in seemingly unchanging monotony such as the shield areas of Canada, Scandinavia and north-west Scotland (Fig.

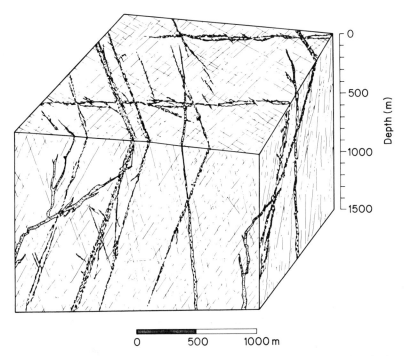

0   500   1000 m

Figure 6.1 Schematic illustration of typical joint and fracture patterns in a body of hard crystalline rocks. Irregular major fractures control the bulk of water movement in the rock (from Nagra, 1985). *Reproduced by permission of Nagra, Switzerland*

6.2). Major fracture zones are usually present every few hundred metres. In fact none of these rocks is homogeneous either in terms of its mineralogical composition, its physical and chemical properties, or its fracture pattern. While some properties remain fairly constant from one area to another, this inhomogeneity must be taken into account in any detailed assessment of rock behaviour.

Basalts (solidified lava) typically occur in flows of variable thickness (up to several hundreds of metres) often layered one on top of another (Fig. 6.3) such that the total thickness of a 'pile' can reach several kilometres and extend over thousands of square kilometres. They display similar jointing and porosity to the other crystalline rocks but have an extra discontinuity to be considered, caused by the discrete nature of the flows. Tops and bottoms of flows are often rubbly or foamy. If long periods have separated basalt lava eruptions, soil layers can form which are subsequently buried under the next flow and persist as subhorizontal bands of higher porosity and permeability within the lava pile. However, since individual units can be some hundreds of metres thick, the potential for siting a repository in basalt clearly exists.

Since all of the cyrstalline rocks have formed at very high temperatures and either crystallized from a molten liquid or recrystallized as a refractory residuum

Figure 6.2 A typical area of ancient crystalline basement rock 'shield' terrain, here with isolated mountains formed of eroded overlying sediments; the north-west highlands of Scotland (courtesy of British Geological Survey)

during high temperature, solid-state metamorphism, it follows that they display considerable thermal stability and present few problems from the viewpoint of heat emitting HLW. Their hydrothermal behaviour, the interaction with warm groundwaters in the repository near-field, can, however, act as a control on near-field chemistry, and was discussed in Chapter 5. They are also reasonably good conductors of heat, have high compressive strengths, and at moderate depths (up to about 1500 m) can be easily and economically excavated to form self-supporting caverns.

At the time of writing, the various national research projects in crystalline rocks (Sweden, France, Finland, Canada, Switzerland and, to a lesser extent, the UK and the USA) have concentrated their efforts almost entirely on granitic or metamorphic rocks. The only work on basalts is the BWIPP (Basalt Waste Isolation Pilot Plant) project at Hanford (Washington State) in the USA. In addition the USA is giving equal consideration to a further 'crystalline' rock type; volcanic tuff, at the Nevada Test Site (NTS). This rock type is essentially compressed air-fall volcanic ash which may have been partly welded by its own heat immediatley on deposition. Vast thicknesses occur in many parts of the world. At NTS it is compact and quite porous, with good engineering and hydrogeological properties. Owing to the presence of zeolite minerals in the pores

Figure 6.3 Bedded lava flows; the basalts of Northern Ireland. The jointed nature of the rock, and the rubbly and discordant nature of the tops and bottoms of succeeding flows can be seen clearly (courtesy of British Geological Survey)

it can also display a high radionuclide sorption capacity, although zeolites generally possess low thermal stability and repository temperatures would probably need to remain below 85°C (Smyth, 1982).

Argillaceous rocks

While the crystalline rocks have fairly restricted physical properties, the clay-rich (argillaceous) rocks can provide a wide variety of geochemical, physical and hydrogeological environments which could be suitable for waste disposal. Unfortunately, the various properties which make this rock type suitable for disposal are often incompatible, and any clay unit is likely to represent a compromise of the potentially useful parameters. For example, the softer plastic clays have negligible permeability, respond to stress without fracturing, and act as good sorbing media for leached radionuclides. However, they occur only at relatively shallow depths, are often associated with more permeable rocks, have low thermal stability, and are difficult to excavate without complex mining and tunnel-support techniques. Conversely, the harder and more massive argillaceous rocks (generally considerably older and having undergone compaction, diagenesis and often low grade burial metamorphism) behave more as crystalline rocks in that they are fractured and less porous, with a lower clay mineral content and hence a lower sorption capacity.

Because argillaceous rocks are nearly always sedimentary in origin they frequently occur as stratified components of very large scale geological structures (Fig. 6.4) and individual beds of clay-rich rocks may extend for many tens of kilometres. Consequently, groundwater movements both within the clay formation and in over- and underlying rocks (sandstones, limestones, chalk, etc.) tend to be on a larger scale than in the case of crystalline rocks; in other words the potential pathlengths of radionuclide migration may be much longer. A highly impermeable clay unit will only allow the passage of significant amounts of water, probably upwards or downwards into adjoining more permeable beds, if the hydraulic gradient across it is relatively high (e.g. artesian conditions). If the gradient is small the advective flow of water could in fact be very much lower than rates of diffusion. As a result, such a unit would not act as a flow conduit along its length. If a repository were situated in it, any leakage would tend to occur into the adjacent units, and it is these which would act as the migration pathway to the biosphere.

Clearly a wide spectrum of properties and flow environments exist, from plastic clays, through more consolidated but still plastic units; compacted mudstones with poorly developed fracturing; shales with strongly oriented fracture planes but remaining relatively soft and plastic; to brittle and highly fissured slates (Fig. 6.5) and phyllites which behave more as crystalline rocks. There is adequate scope within this category to tailor the disposal system to a particular environment, for example by reducing the thermal loading in the case of HLW. This style of approach is evident in the national research programmes.

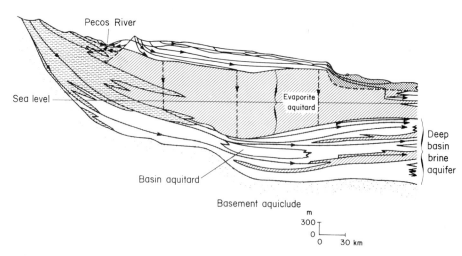

Figure 6.4 Schematic cross-section of a typical sequence of sediments in a deep sedimentary basin, showing directions of groundwater movement (after NEA, 1984a). *Reproduced by permission of OECD*

Figure 6.5 Strongly jointed and fractured slates; argillaceous rocks in their most compacted state where they behave hydraulically in essentially the same way as fractured crystalline rocks (courtesy of British Geological Survey)

Belgium now has considerable experience in studying plastic clays for HLW disposal and has mastered most of the technical problems involved in working in these rocks while taking advantage of the undoubted benefits. Italy, the UK and the USA are putting effort into somewhat more consolidated units but, on the whole, clays are receiving much less attention at present that either crystalline rocks or salt, at least as far as deep disposal is concerned.

Evaporites

Evaporites are units of soluble salts of Na, Mg, K and Ca, formed in the geological past by evaporation of shallow lagoons and lakes in cyclic episodes. Considerable thicknesses of simple and complex chlorides, sulphates, carbonates, and so on may be laid down in layers with various degrees of mixing of salts, and subsequently buried as part of a stratified sedimentary sequence. As part of the process of burial they have often recrystallized to simpler and more stable salts. The commonest salt is halite, rock-salt (NaCl), although it is generally accompanied by many of the 80 or so other principal evaporitic minerals. Despite the frequent inhomogeneities, very thick units of almost pure halite (and several of the other salts) can be found (Fig. 6.6). Because of its plastic nature halite is very susceptible to tectonic stresses deep in the earth and can be mobilized *en masse* due to its own density disequilibrium (being less dense than the surrounding rocks), rising in inverted tear-shaped blobs known as diapirs (Fig. 6.7). Halokinesis, as this phenomenon is known, is not universal in regard to salt deposits; for example there are no diapiric (or 'salt-dome') structures below the UK mainland (although there are extensive simple bedded evaporite units) nor does it always result in the salt domes reaching the earth's surface. Several of the widespread continental European and US Gulf States salt domes show no evidence of contemporary movement, while others are rising very slowly, at the rate of a few millimetres a year. The time scales for such movement are so long that they present no particular problems from the waste-disposal viewpoint.

Dome salts appear to be thermally more stable than their simple bedded counterparts. Salt is an excellent conductor and disperser of heat and is also, to

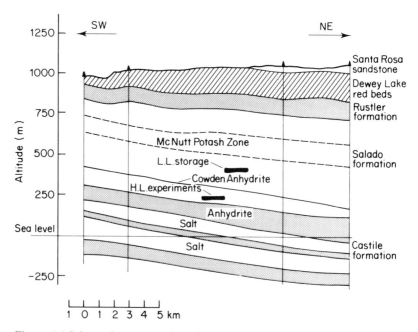

Figure 6.6 Schematic cross-section of the WIPP site in New Mexico, USA, for disposal of long-lived defence wastes in a bedded salt formation (Salado formation)

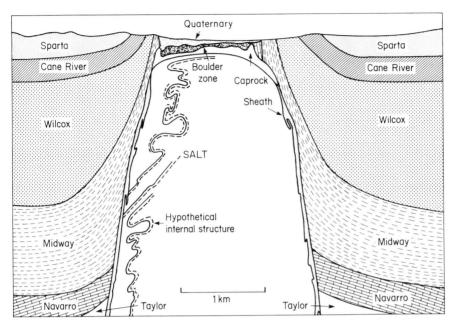

Figure 6.7 Cross-section of a typical salt dome (Rayburn's dome in Louisiana, USA), showing the contorted structure of originally horizontal evaporite beds within the dome, and the surrounding sediments upthrust during the rise of the dome. The presence of Quaternary sediments on the top of the dome indicates that it has been exposed at the surface within the last few million years. *Reproduced by permission of OECD*

all intents and purposes, impermeable to water, while containing only small amounts itself. The historic lead that salt has acquired in the HLW disposal field reflects the emphasis that was placed on an impermeable host rock in the earliest days of the research. Despite its obvious solubility in freshwater, salt is extremely stable in the geological environment, contains few if any fractures, is self-sealing owing to its simple plastic response to stress, and can be mined with ease. The consequent problems which surely accompany such an apparently ideal medium are: the frequent and often unpredictable occurrences of highly unstable mineral bands such as carnallite which can decompose at low temperatures to give out large quanitities of water, the presence of fluid inclusions which can migrate towards a heat source carrying corrosive mixed brine with them, and the risk of flooding an operating repository, which is essentially a dry mine operating well below the water table present in overlying sedimentary units, some of which might be important aquifers.

An experimental underground facility in a bedded salt was in operation as long ago as 1967 in the 'Salt Vault' project at Lyons in Kansas. Since then, and after a long pause in the work, the USA has committed much more effort to this rock type with the WIPP (Waste Isolation Pilot Plant) project in New Mexico, the Avery Island *in situ* heater experiments in the Gulf Coast (Louisiana), and characterization of a potential repository site in Deaf Smith County, Texas. At the same time the Federal Republic of Germany has put considerable emphasis on the use of dome salt for storage and possibly disposal of intermediate level wastes at the Asse mine, and is presently involved in the exploration and development of the salt dome beneath Gorleben for high- and intermediate-level waste disposal. With the phasing of the current commercial waste programme in the USA, the additional WIPP commitment, and the entry of both the Netherlands and Denmark into the international salt programme, it appears that this option might well be among the first to be pressed through to an operating conclusion.

FACTORS CONTROLLING MIGRATION THROUGH THE GEOLOGICAL BARRIER

Radionuclides can leave the near-field environment by either advection (entrainment in moving water) or diffusion (dominant in very slow moving or static pore waters). They may be in solution in groundwater, in particulate form (as colloids and suspensions) or sorbed on to other suspended material. The speciation behaviour of many of the significant radionuclides, in particular the actinides, is extremely complex (see Chapter 5). The principal controlling factor is the nature of the geochemical environment, dominated by the groundwater chemistry. For the present however, we will ignore chemistry and concentrate on the physical properties of the groundwater environment.

Hydrogeology of permeable rocks

The 'water table' is the level in the ground below which water saturates the void space or porosity of a rock, which is made up of fissures, fractures and pores. Even above the water table the pores may be almost saturated with percolating water. In most temperate regions of the earth the water table is within a few metres or tens of metres of the surface. In the case of many argillaceous and crystalline rocks in the UK, where there is high rainfall and modest relief, the water table is to all intents and purposes coincident with the surface, particularly in flat lying areas (Black and Chapman, 1981). In all but exceptional cases in arid desert environments (such as the Nevada Test Site, NTS) we can say that at proposed repository depths the rock will be saturated with water. At NTS the water table is many hundreds of metres deep, and a repository might be situated in the unsaturated (but still wet) zone above it.

Groundwater moves in response to variations in 'head' from one point to another. It can also be driven through the rock by the thermal effects of the waste which produce density differences (buoyancy) in the water. Under natural conditions head differences are most frequently produced by topographic effects—simple height difference (Fig. 6.8), giving rise to hydraulic pressure gradients down which water moves. They can also be produced by the presence of zones or units of rock with different hydraulic conductivity (the ability to transmit water; broadly equivalent to electrical conductivity).

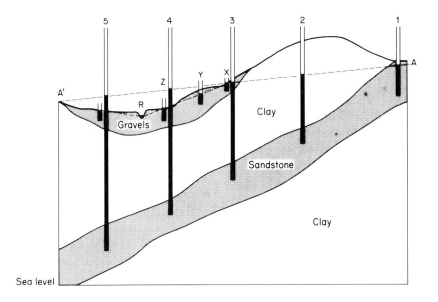

Figure 6.8 Simplified, and highly schematic illustration of hydraulic 'head'. If pipes were sunk into various points in the sandstone aquifer, then water would rise in them to the levels indicated in response to the head at each point. The hydraulic gradient, 'A–A', controls the direction and rate of groundwater flow in the aquifer; water moves down gradient. Pipes 1–3 could represent ordinary water wells, whereas 4–5 demonstrate what are often loosely referred to as 'artesian' conditions, in a confined part of the aquifer. Wells at these points would overflow at the ground surface. The heads in the river (R) valley gravels (pipes X–Z) are close to the surface, and in this case the line showing the hydraulic gradient in this aquifer unit also represents the water table, and the closely stippled area above it is the unsaturated zone. None of these wells overflows. To the left of pipe 3, the heads in the sandstone are higher than those in the gravels, and there is consequently a vertical hydraulic gradient which would allow very slow seepage of groundwater from the sandstone, upwards through the intervening clay, into the gravels. If heads in the upper clay formation were higher than those in the aquifers, then seepage of clay groundwaters might occur both upwards and downwards into the gravels and sandstones. The importance of knowing the heads through such a series of formations in order to predict directions and rates of groundwater movement is clear. Head is usually measured in metres above sea-level

The hydraulic head can vary considerably in different regions of a thick sequence of sediments, or within different fracture zones of a crystalline rock body, with little or no obvious relationship to the depth below the surface. This indicates that there may be very poor hydraulic connection between these different zones caused by intervening regions of low hydraulic conductivity (generally referred to as 'low permeability') rock. High pressures cannot easily be dissipated through such regions, as the water movement required to transfer energy is very slow. Zones of elevated pressure may result from long-distance connection via high-permeability rock formations to topographically higher areas. They can also be caused by differential compaction processes, or the presence of hydrocarbons in sediments. As a consequence, quite complex patterns of head can occur within a given volume of rock, resulting in flow patterns which might not easily be predicted from first principles. Artificial pressure differences can be caused by pumping an operating repository to keep it dry, thus locally reducing the ambient water pressure and creating a pressure gradient towards the pump, drawing water into the repository. Depending on the hydraulic properties of the rock, these pressure perturbations can persist for some time after pumping has stopped.

The hydraulic conductivity of a rock is a function of the size and degree of interconnection of its pores and fractures. The smaller and more tortuous the structure of the void space, and the greater the surface effects in the pores, then the lower is the hydraulic conductivity.

These parameters of hydraulic conductivity and head gradient are linked in the well known equation which defines the simplest form of Darcy's Law of flow through porous media (Fig. 6.9).

$$Q = KIA$$

where Q = volumetric rate of flow (m^3/s)
K = hydraulic conductivity (m/s)
I = hydraulic head gradient (m/m)
A = cross-sectional area through which flow occurs (m^2)

There is some controversy as to the degree of applicability of Darcy's Law to fractured rocks and other rocks with very low hydraulic conductivities, such as plastic clays. It is basically a means of representing flow through an isotropic porous medium like a sponge, but is widely used in the treatment of fractured rocks where there are clearly varying degrees of anisotropy in hydraulic properties caused by the very nature of the fracture patterns themselves. It is thought, however, that for fractured rocks Darcyan behaviour is a reasonable assumption if a large enough volume of rock is considered, and attempts are presently being made to test this approach against network flow models which more accurately represent anisotropic behaviour.

The Darcy equation is very similar to the equation linking electrical current, resistance and voltage. In the form above it provides a volumetric flow rate, but no velocity of flow from one point to another down the hydraulic gradient. To obtain this simply for unit cross-sectional area (A), we divide Q by θ, the 'effective' or 'kinematic' porosity of the rock, or that part of the total void space which takes part in flow. In a fractured rock this essentially means the fractures only, and not the intergranular pore space.

Early work on fractured rock hydrogeology tended to treat the rock as a whole, and either looked at homogenized hydraulic properties on a large scale or averaged them. It is now clear that the hydraulics of the very low permeability rock between *major* fractures (i.e. blocks of rock tens or hundreds of metres in size) is of importance only to near-field modelling. Far-field transport will occur

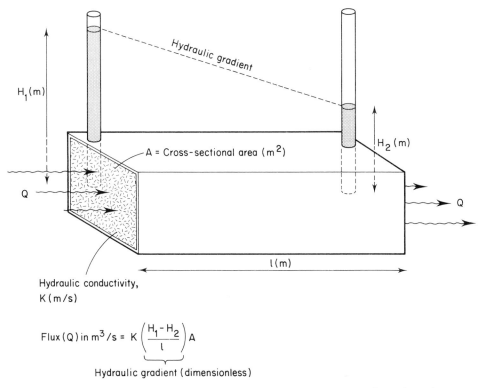

Figure 6.9 Schematic representation of Darcy's Law of groundwater flow through a confined volume of porous sediments

in major fractures (those which occur perhaps every few hundred metres, may connect to the surface, and have high porosities and hydraulic conductivities), linked by smaller hydraulically active fractures to the bulk rock 'blocks'. Within any of these fractures preferential pathways exist for water movement (channeling), so they should not be considered as planar features (e.g. the 'parallel plate' model of equal fissure aperture all across the plane), but rather as planes within which 'worm tubes' lie (Fig. 6.10). This is because fissures have variable apertures owing to the rough nature of their surfaces and differences in the properties of their infill minerals. Models for fractured rock hydrogeology have moved from average porous medium equivalents, to single fracture dominated models connecting porous medium blocks, to probabilistic fracture network models, and now to random preferential paths. There is some way to go yet.

It is at this point that we can begin to see more clearly the difference between fractured rocks and unfractured 'porous' rocks such as clay-rich sediments. The latter often possess very high porosities, but if they lack fractures, may have no well-defined preferential flow routes. In this case all flow must be through the total porosity, although this may be anisotropic, owing to silty bands, for

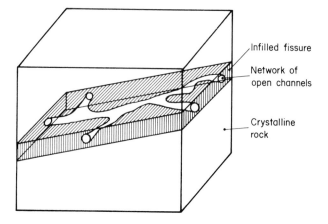

Infilled fissure

Network of
open channels

Crystalline
rock

Figure 6.10 Illustration of the concept of flow channelling
through higher conductivity zones in a planar fracture

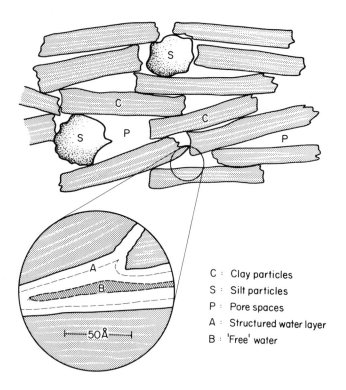

C : Clay particles
S : Silt particles
P : Pore spaces
A : Structured water layer
B : 'Free' water

Figure 6.11 Highly schematic depiction of compacted clay porosity. Silt particles prop
open some quite large pores, but the majority of pore spaces are between roughly aligned
laminar clay particles, with a maximum pore size of about 100 Ångstroms (10 nm). Within
these pores the water 'double-layers', which are structured and bound to the clay surfaces
by electrostatic forces, almost touch, leaving little free water. The charge pattern at the
clay surfaces can concentrate various cations, but tends to exclude anions

example. If we took a piece of unfractured sandstone, which has a very high porosity and relatively large and open pores, the hydraulic conductivity, or ability to transmit water, is quite high. However, a plastic clay, while also possessing a high porosity, has very small pores which are interconnected in an extremely tortuous manner. In many cases only a very small proportion (perhaps less than 10 per cent) of the water in these pores is actually free to move (Henrion, et al., 1985). The remainder is physically bound to the surface of the clay minerals. Thus while a plastic clay may have a total porosity of 30 per cent, its effective 'flow' porosity may be much less than 5 per cent. (Fig. 6.11). In addition the hydraulic conductivity is exceptionally low, and these types of rock are often considered to all intents and purposes as impermeable. While flow through an open-pored sandstone would be very close to simple Darcyan behaviour, the flow through plastic clays of very low permeability, as with low hydraulic conductivity fractured rocks, is rather more difficult to model. In the case of clay, flow under the hydrogeological circumstances suited to a repository site (low hydraulic gradients) is so slow that we begin to consider time scales and rates which are comparable to molecular diffusion processes. Thus, in clays, the initial part of the migration process takes place by diffusion, with possible subsequent advective transport in flowing groundwaters in overlying sediments.

Table 6.1 lists some representative values of the parameters discussed above, and the extreme ranges of values can be seen immediately. It is clearly possible, within several of the rock types, to choose values of the various parameters which will give rise either to very slow or very rapid transit times of groundwater from

Table 6.1 Likely maximum and minimum values for the principal hydraulic properties of various sediments and crystalline rocks, in typical environments which might be considered for disposal purposes

Rock type and depth	Hydr. cond. m/s	Porosity	Gradient	Flux lit/yr/m^2	Velocity m/yr
Plastic clay					
Above 100 m	10^{-7}	0.50	0.2	640	1.28
	10^{-10}	0.30	0.05	1.6	0.00053
Below 100 m	10^{-8}	0.50	0.2	64	0.128
	10^{-12}	0.30	0.05	0.0016	5.3×10^{-6}
Shale/mudstone					
Above 100 m	10^{-6}	0.30	0.2	6400	21.3
	10^{-9}	0.20	0.05	1.6	0.008
Below 100 m	10^{-7}	0.25	0.2	640	2.56
	10^{-10}	0.05	0.05	0.16	0.0032
Crystallines					
Above 100 m	10^{-7}	0.05	0.1	320	6.4
	10^{-9}	0.01	0.001	0.032	0.0032
Below 100 m	10^{-8}	0.01	0.1	32	3.2
	10^{-11}	0.001	0.001	0.00032	0.00032
Aquifer	10^{-4}	0.10	0.01	32000	320
	10^{-7}	0.05	0.0005	1.6	0.032

114

one point to another. The reality of such scenarios will be discussed in more detail later.

We have thus defined a group of very simple *hydraulic* parameters which are needed for the radionuclide transport model. These can be used to quantify groundwater flow rate, and hence to a first approximation the maximum transport rate of the released radionuclides. They do not, however, define what is probably the most problematic feature, the length and direction of the potential flow and migration paths; the *path* and *pathlength*. In this respect there is a

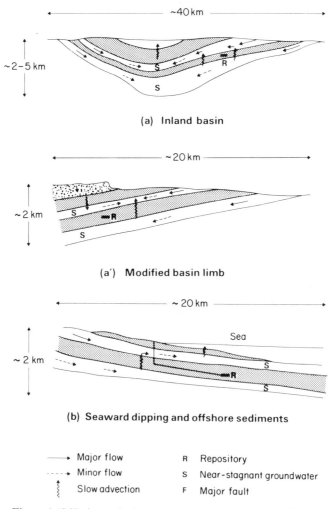

(a) Inland basin

(a') Modified basin limb

(b) Seaward dipping and offshore sediments

⟶ Major flow		R	Repository
----→ Minor flow		S	Near-stagnant groundwater
Slow advection		F	Major fault

Figure 6.12 Hydrogeological environments in the United Kingdom thought suitable for deep disposal of long-lived wastes (after Chapman *et al*, 1986b). Similar environments are common throughout northern Europe. The darker formations in the sedimentary systems (a-b) represent low hydraulic conductivity

considerable difference between massive bodies of crystalline rock and stratified units of sedimentary rock. In both cases it is likely that large scale structural features, such as major faults, will act together with topographic features as boundaries to flow units or '*flow cells*'. However, in the case of crystalline rocks current thinking centres upon the idea of very frequent flow boundaries giving rise to many small-scale localized flow cells within a body of rock, rather than the extensive regional flow patterns found in stratified formations. These concepts are portrayed schematically in Fig. 6.12. In all cases the shortest pathlength to the

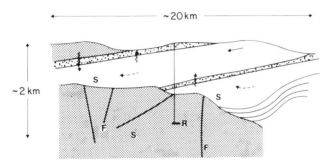

(c) Basement rocks under sedimentary cover

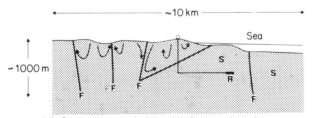

(d) Crystalline rock; low relief coastal environment

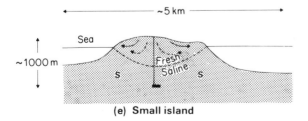

(e) Small island

units, such as clays, mudstones, indurated sandstones, bedded evaporites, etc. General directions of groundwater movement, together with appropriate repository locations, are indicated. *Reproduced by permission of the International Atomic Energy Agency*

biosphere is directly upwards from the repository to the surface. This is an extremely improbable escape route for the waste as few hydraulic conditions can be envisaged which would give rise to unmodified vertical flow. It can be seen from Fig. 6.13 that the possibility exists for some quite short pathlengths in certain geological environments. The flowlines depicted (by two-dimensional finite element modelling) are in response to typical predicted hydraulic gradients. In a formation which displays decreasing hydraulic conductivity with depth, the greater the depth to which a flowline penetrates, then the smaller is the volume of water moving along it and the slower the velocity. In other words, a near-surface flowline might represent large volume transport over perhaps a few hundred years, whereas a deep line would depict a very small volume of water taking perhaps tens or hundreds of thousands of years to re-emerge.

In the examples shown in Fig. 6.12 for stratified sequences, the pathlengths can be considerable, and time scales may be comparably long. It should also be borne in mind that, in a stratified sequence, leakage of the contaminated groundwater may take place out of the repository host formation, and subsequent transport may be in an entirely different underlying or overlying rock type with different hydraulic properties. Because of the variability in hydraulic properties of sedimentary sequences, anomalous pressure differences and flow patterns can develop which may drive water up or down through the host-rock unit (Black and Barker, 1981a; Black et al., 1985). In some environments, locations can be chosen where any groundwater flow must take place in a unit which will conduct contaminated water under the sea before discharge to the biosphere, hence considerably diluting the release. Such environments can be found in both sedimentary and crystalline rocks (Chapman, et al., 1986). The logical extension of this idea is to select a small island as a repository site, where all discharge must of necessity be to the sea. Some island environments have a unique hydrogeological feature known as the 'Herzberg lens', which is the phenomenon of freshwater in the rock pores 'floating' on top of denser saline groundwater at depth (see Fig. 6.12). The depth to the saline/freshwater interface is controlled by the thickness of the freshwater lens (reflecting the island's topography and infiltration rate/rock permeability) and mass exchange across it is limited. It is thought that groundwater movement just below the interface (in the saline waters) is virtually non-existent and would thus prove a potentially ideal site for a repository. Sites on the coast of the mainland possess some of the advantages of an island and are also well worth considering.

The fundamental hydrogeological properties of permeable rocks required for simple transport modelling are thus: hydraulic conductivity; kinematic (flow) porosity; regional and local hydraulic gradients; geological features controlling flowpaths. These factors may vary quite considerably even within a given repository site. When performance modelling is carried out to assess a potential deep-disposal site a much wider area of ground than the locality of the repository itself must be considered. This is especially true of stratified units where data on these properties must be available along what might be quite considerable

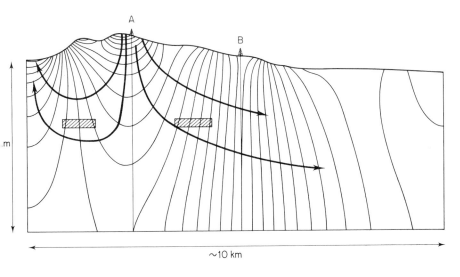

Figure 6.13 Typical results of a very simple two-dimensional finite element model of groundwater flow. A block of crystalline rock, with zero-flow boundaries assumed at the base and sides, and hydraulic conductivity decreasing progressively with depth. The curves are equipotentials (simply, lines of equal groundwater head), which can be seen to be controlled by the topography. Groundwater would flow down a potential gradient, that is at right angles to these equipotentials. A borehole at A would encounter progressively decreasing heads with depth, while one at B would find similar head values throughout the whole borehole. Some possible flow paths are shown. Flow volumes and velocities will decrease markedly with depth. The repository situated on the right appears to be in a better position than the one on the left, as pathlengths are potentially both deeper and longer, and return of activity to the surface consequently less likely

potential pathways. Some idea of the variability of parameter values within any given body of rock was given in Table 6.2. In crystalline rocks the hydraulic conductivity is thought to decrease markedly with depth (Fig. 6.14). In regions which have been glaciated and subjected to permafrost to considerable depths, the effects of loading and unloading stresses caused by the presence of hundreds or thousands of metres of ice overburden, have caused fracture enhancement and consequent increased permeability in the upper few hundred metres.

This short list only defines the Darcian parameters in their simplest form. To obtain a true understanding of anisotropic flow behaviour the equations become more complex and require further parameters to be determined. Some of these parameters are essential derivatives of field experimental techniques which endeavour to determine values of anisotropic hydraulic conductivity, storativity, dispersivity, transmissivity and other factors. These detailed hydrogeological factors are outside the scope of this book, and the reader is referred to specialized texts such as Freeze and Cherry (1979) or de Marsily (1986).

118

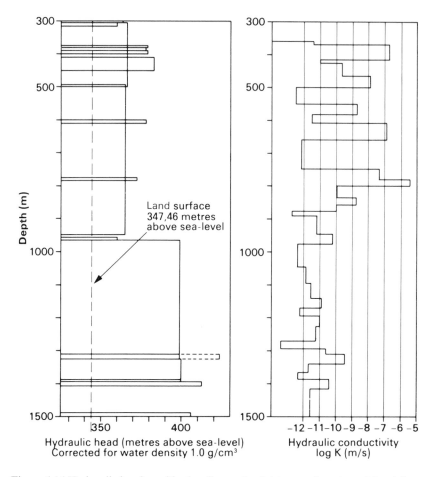

Figure 6.14 Hydraulic heads and hydraulic conductivities as a function of depth in the granitic basement of northern Switzerland. The progressive decrease in bulk conductivity is clear (from Nagra, 1985). *Reproduced by permission of Nagra, Switzerland*

Modelling groundwater movement

The simplest mathematical models of groundwater movement are analytical solutions of the flow equations which assume uniform rock and hydraulic properties. For example, if we were examining flow through silty sands in a river flood plain, which had a hydraulic conductivity of 10^{-4} m/sec, a porosity of 40 per cent and a hydraulic gradient of 1 in 100, then the simple application of Darcy's Law would give a flow velocity of about 20 cm/day. In reality the hydraulic properties of a groundwater flow path are often variable and this simple calculation provides no information about the three-dimensional flow of water and about different volumes of flow at different depths, or the effects of recharge and discharge of the aquifer. More complicated analytical solutions can take some of these factors into account, but more sophisticated techniques are available. The most commonly used method, which is now being applied to problems of radionuclide transport in groundwaters, is finite element (FE) modelling. This technique is complex and generally requires main-frame computer power. Essentially a finite element model is a 2- or 3-D (dimensional) representation of a volume of rock which is divided into boxes (elements) in a complex grid pattern (Fig. 6.15). The physical and hydraulic properties of each element can be defined separately, so any variations resulting from different rock types, faults and discontinuities, and layered sequences can be built into the model. The sizes of the elements can be varied across the grid to obtain more detail in specific zones. In this way factors such a anisotropy in hydraulic conductivity can be represented accurately. The whole grid must be defined by boundaries, usually lines across which no flow can take place.

Computer programs can calculate precise hydraulic gradients (by defining lines of equal hydraulic pressure) and flow pathlines and volumes of flow by taking each element and calculating the physical effect of the surrounding elements. A commonly used FE-model in the UK is called NAMMU (Rae *et al.*, 1981) and typical results were shown in Fig. 6.13. In its basic form this 3-D model examines flow caused by topographically defined hydraulic gradients in the rock. However, as discussed earlier, in the case of disposal of heat-emitting HLW, flow may also be driven by thermal buoyancy or convection. NAMMU can also model these types of groundwater movement by examining thermal profiles around a repository and calculating the relative expansion of the rock and the water. Regional topographic hydraulic gradients can be superimposed on thermal effects to give a picture of groundwater movement at any instant in the thermal life of a repository. From this type of modelling, it appears that convection in a fractured rock can be an important mechanism for groundwater movement, even long after repository temperatures have declined (Hodgkinson, 1980). This is because the driving force depends on the total quantities of heat released, which when integrated over the long periods of time after repository temperatures have become quite low, are still very large (Bourke and Robinson, 1981). Similar models are equally applicable to all types of permeable rock, although different results are obtained in each case (Rae *et al.*, 1983). For shallower burial depths the volumes of rock modelled are smaller and particular attention must be paid to infiltration rates and water movement in the unsaturated zone above the water table. In all cases, the boundary conditions for the models must be chosen with care. Groundwater divides (such as streams, watersheds, impermeable faults, and so on) are usually chosen as being zero flow lateral boundaries, and some simplifying assumption may be made about essentially zero permeability at depth as the lower boundary.

We now have an outline of the ground rules for estimating groundwater transport rates and directions. At this stage we are still assuming that the nuclides do not interact with the rock, and move at approximately the same speed as the water independent of whether they are in solution or suspension. The next

120

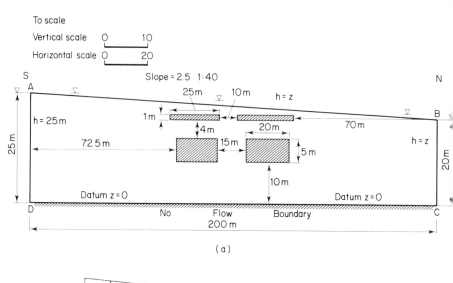

To scale

Vertical scale $\quad$ 0 $\qquad$ 10

Horizontal scale $\quad$ 0 $\qquad$ 20

S

A

Slope = 2.5   1:40

25 m

10 m

h = z

h = 25 m

1 m

4 m

20 m

70 m

B

25 m

72.5 m

15 m

5 m

h = z

20 m

10 m

Datum z = 0

Datum z = 0

D

No   Flow   Boundary

C

200 m

(a)

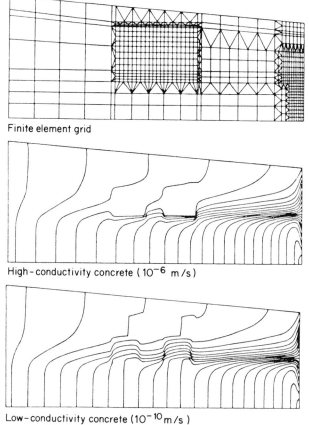

Finite element grid

High - conductivity concrete (10^{-6} m/s)

Low - conductivity concrete (10^{-10} m/s)

(b)

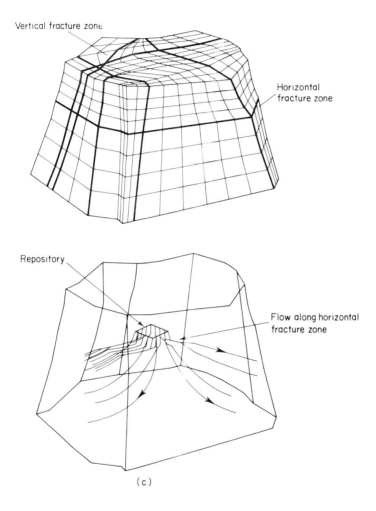

Figure 6.15 A test model (a) for the NAMMU finite element groundwater flow computer code (from the HYDROCOIN exercise; see Herbert, 1985). Two near-surface disposal trenches, constructed of concrete, with a concrete 'anti-intruder' shield above (see Figure 9.2). The finite element grid constructed to model flow in the trench zone is shown in (b), together with the hydraulic head equipotentials calculated by the code for two values of hydraulic conductivity of concrete. A similar exercise is shown for a much larger scale problem in 3-dimensions in (c), for a deep repository in fractured rock (after Herbert *et al.*, 1984)

feature which must be considered comprises the various physical mechanisms which come into play to retard this movement.

Physical dispersion and diffusive retardation

As water flows through a rock, any chemical component dissolved in it might be expected to travel at the same speed as the water. If we take a single tube-like pathway of 'clean' water and inject it at some point with a short pulse of a tracer chemical, then we would expect to see that pulse passing a downstream monitoring point as a spike on a concentration vs time graph. The width of the spike would correspond to the time taken to inject the pulse initially (Fig. 6.16). If the water movement is very slow, however, diffusion of the component into the surrounding water behind and in front of the pulse will become noticeable, the width of the spike will increase, and its height (maximum concentration) will decrease a little. Different types of flow within the pathway (for example turbulent or laminar flow) and frictional edge effects will also give a similar result. These may be quite marked over the extensive pathlengths described in the previous section, at the slow rates anticipated. They will, however, be relatively small when compared with the effect of physical dispersion. This takes place

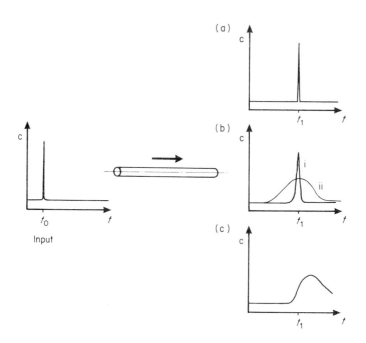

Figure 6.16 Schematic illustration of the movement of a tracer in water flowing along a tubular channel. At t_0 the tracer is injected as a 'spike', shown on the left hand concentration v. time graph. With ideal flow it emerges (a) as an identical spike at t_1. With varying degrees of hydrodynamic dispersion (i and ii) the spike is reduced and blurred (b). If sorption also occurs (c) then the spike may develop a long tail

during flow through a *network* of fissures, or a complex interconnected pore structure, which provide numerous alternative paths between two points, all of different degrees of tortuosity, and hence of length. From first principles, the arrival of the tracer should be at the same time as in the single channel case, as this still represents the shortest route, but now the spike will tail out over a long period of time due to the 'late arrivals' from the longer paths (see Fig. 6.16 again). This phenomenon, known as dispersion, is clearly very dependent on the fracture pattern or pore structure, and the degree of isotropy of hydraulic parameters within the volume of rock being considered. Its net effect is to retard the arrival of the peak of a 'pulse' of radionuclides released into groundwater, and to decrease considerably the concentration in which they arrive at the outflow or measuring point. Noy and Holmes (1986) give a concise description of dispersion: Put simply, dispersion results from mechanical (caused by flow) and physico-chemical (molecular diffusion) mechanisms, due to properties of both the fluid and the material in which the flow is occurring. In general the 'dispersion coefficient' is a symmetric tensor quantity having three independent parameters in the two-dimensional case and six parameters in the three-dimensional case. However, if the material is isotropic and the dispersion is assumed to be proportional to the flow velocity then the mechanical component of the dispersion coefficient can be written in terms of just two parameters, the *longitudinal* and *transverse dispersivities*. Bear (1972) shows how the components of the dispersion tensor can be written in terms of the dispersivities, the components of fluid velocity, and the molecular diffusion. The properties of some rocks, such as those in which flow occurs in a few major fractures, result in low values of dispersion so that tracer concentrations remain high over large distances. Others, such as poorly consolidated sandstones, have higher dispersivities and tracers become quickly diluted.

Recently attempts have been made to model combined flow and dispersion in fractured rocks using a statistical approach to the fracture network (Robinson, 1983). This technique constructs a probabilistic model of fracture pattern and interconnections using as a basis real statistical field data, and then uses percolation theory to calculate water movement through the model network. This approach is likely to be the most feasible way forward since a full *in situ* determination of a fracture system is not likely to be possible (or tractable) in practice. In principle this is a significant step towards bridging the gap between purely porous medium flow models and simple-fracture flow models. At present, results indicate that the former type of model is valid if the volumetric scale is large compared to the fracture intervals. Hydrodynamic dispersion is complemented by chemical retardation (described in the next section), and in fractured rocks by a further phenomenon called 'dead-end pore diffusion' or *matrix diffusion*.

In the discussion of the hydraulic properties of rocks, two styles of porosity were described. The effective porosity is what we have been dealing with until now, that is the active porosity which takes part in advective flow. In a fractured rock however, the effective porosity might amount to only 0.1 per cent of the rock by volume, whereas the total porosity may be up to 2 per cent or more. In this case only 5 per cent of the total void space in the rock is 'flowing', the remaining 95 per cent is filled with water which to all intents is static. The water flowing along a fissure system is in contact with the large reservoir of dead water, and any components in the moving water (e.g. dissolved radionuclides, colloids) will be free to diffuse down concentration gradients into the main body of unfractured rock, subject only to constraints set by the pore dimensions (Fig. 6.17). As they diffuse out into the dead pore space, the situation is analogous to a man stepping off a jungle track into the trees. He will easily get lost in the complex undergrowth and may remain there forever. He might even be eaten by a tiger—but more about sorption later! In the case of a chemical diffusing into static water, there is no mechanism other than diffusion to return components to the flowing stream, and this can only take place when the chemical gradient is reversed and concentration in the rock pore-waters exceeds that in the fissure water (i.e. when the main pulse has passed).

There has been much effort in the study of this phenomenon, and initial calculations (e.g. Neretnieks, 1980; Grisak and Pickens, 1980; Barker, 1982) indicate that this could be an extremely effective mechanism of retardation by physical means alone. If we also give credit to chemical reaction of the nuclides with pore walls in the dead space, this retardation mechanism is potentially so efficient that most if not all, radionuclides decay to insignificant levels before they are released to the biosphere. As yet, the existence of long-distance networks of connected pores extending from fracture zones throughout the entire matrix has not been proven for relevant rock types. The potential importance of the mechanism is such, however, that it is under extensive investigation in laboratory, field and natural analogue studies.

In an unfractured rock this particular retardation mechanism may not occur unless similar preferential flow paths exist. Instead diffusion takes place throughout the pore structure of the rock. However, since an 'effective porosity'

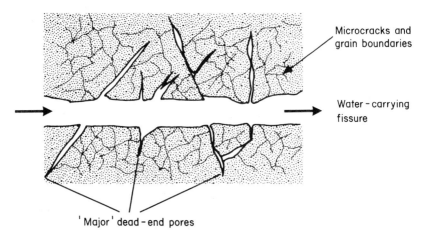

Figure 6.17 Microcracks and dead-end pores which may permit matrix diffusion in fractured rocks (see text)

can also exist in clays, the situation is complicated if any noticeable flow does occur (under high hydraulic gradients for example). Since the non-mobile water is present in the same pore space as the mobile water, chemical exchange by diffusion can take place between them. Owing to the high surface charge on clays however, anions may be excluded from the non-mobile water.

An additional, poorly understood, process, known as surface diffusion may also be operative, whereby cations migrate more quickly in the non-mobile water layers at highly charged mineral surfaces (Rasmuson and Neretnieks, 1983). This rather complicated system has not been fully assessed, and for most modelling purposes it is usually assumed that free diffusion can occur throughout the whole available porosity. Any components of advection (flow) are superimposed on this, although under the majority of circumstances the rate is much slower than for diffusion.

The dispersion and diffusion effects described in this section can be experimentally quantified, as could the more basic flow properties of the geological barrier described earlier.

Chemical interaction of radionuclides with the rock

As radionuclides migrate from the immediate environment of the waste they are affected by interaction, not only with the corroded package and buffer material, but also with the massive body of rock through which they must subsequently pass, and which acts as the principal chemical buffer to all solid-fluid interactions. The rock buffers the composition of the groundwater and limits its rate of supply and thus, as noted in Chapter 5, controls the way in which the near-field performs. Radionuclides diffusing out of the buffer into the host rock will either be faced with a continuous static body of pore water (e.g. in a clay), in which case diffusion continues out into the far-field, or they will be entrained in slowly moving groundwaters in a pore or fissure system (e.g. in a crystalline rock). In both cases a contamination 'front' will develop around individual packages and slowly increase in volume, being shaped by flow or diffusion anisotropy, and by interaction with similar fronts from adjacent leaking packages. Rather complex concentration patterns might occur over very long time periods within the volume of a repository. In some scenarios, we might consider the repository as a whole being a source for subsequent far-field migration, although there is little doubt that the majority of waste components will be fixed or will decay to negligible levels within the actual repository volume, or even in the immediate neighbourhood of their source package, as shown in Chapter 5.

As radionuclides pass from the near-field engineered barriers to the far-field, the progressive change in water chemistry will cause the chemical form of some elements in solution (speciation) to alter, which may be reflected by changes in solubility or extent of retardation. If a very sharp boundary exists between different chemical conditions in these regions, extensive precipitation of radionuclides associated with secondary mineral formation may occur at this interface. As mentioned in Chapter 5, such precipitation could act as a source of colloidal material but it may also block pores and thus decrease the permeability of some low porosity rocks. The thermodynamic modelling approach used to evaluate solubility in the near-field is also applied to the far-field with the slight advantage in the latter case that the chemistry is usually better defined. As the migration path of the radionuclides traverses different rock environments, groundwater chemistry will slowly evolve with consequent changes in speciation. At present levels of sophistication, this could only be modelled as a stepwise transition and only the most important groundwater characteristics (redox conditions, pH, temperature, concentrations of major ions) could be taken into account.

The most important processes of radionuclide/rock interaction in the far-field are generally grouped together under the loose heading of 'sorption', although few of them would be included in a physical chemist's understanding of this term. The mechanisms involved are illustrated in Fig. 6.18 and include:

(a) purely physical processes which retard migration such as molecular filtration, ion exclusion and diffusion into dead-end pores;

(b) direct chemical reaction with rock surfaces involving physical adsorption, chemical adsorption or direct incorporation (mineralization) into the rock structure;

(c) indirect chemical reactions e.g. precipitation caused by enhanced concentrations at the rock surface.

In principle, most of these processes are applicable to suspended or colloidal material in addition to that in true solution. The inclusion of the category of physical processes may seem incongruous here, especially as they have been considered in the previous non-chemical section. The magnitude of these processes can, however, be influenced very greatly by radionuclide speciation (especially if formation of very large or charged complexes is possible) and, in practice, they are often impossible to quantify experimentally in isolation from chemical sorption. This will be discussed further in Chapter 7.

As mentioned in the previous chapter, sorption is usually described by a simple partition constant (K_d) which is the ratio of rock phase radionuclide concentration to that in solution in the groundwater. Use of this simple approach in migration modelling requires the assumption that sorption is reversible, reaches equilibrium quickly, is independent of variations in water chemistry or mineralogy in particular regions of the flow path, and is not affected by changes in rock water ratio, or in the concentration of the species involved or other solution components which are being sorbed as they migrate. Such assumptions rarely hold in reality but, in some cases it can be shown that this approach overestimates the extent of migration (i.e. is conservative) with appropriate choice of K_d values. More sophisticated migration models separate out several different retardation mechanisms, and may also include the concentration dependence of chemical sorption by using empirically derived functions, often

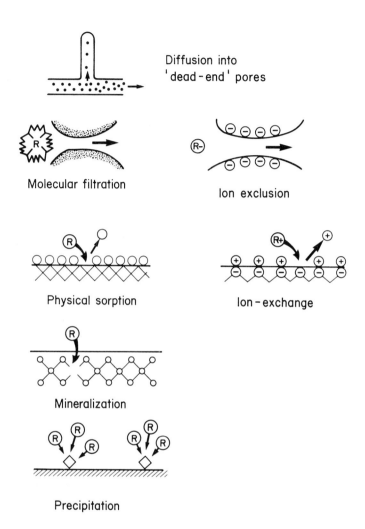

Diffusion into 'dead-end' pores

Molecular filtration

Ion exclusion

Physical sorption

Ion-exchange

Mineralization

Precipitation

Figure 6.18 Schematic representation of the many possible retardation mechanisms possible as a result of interaction between a rock surface and radionuclides either in solution or in particulate or colloidal form. 'R' represents the radionuclide species (from McKinley and Haderman, 1984). *Reproduced by permission of Nagra, Switzerland*

called *isotherms*. The complexities of the resulting calculations (and lack of data) have, as yet, limited the application of such models.

Present safety assessment models generally use a simple empirical K_d approach and, on the basis of the discussion above, this would seem to be justifiable. Problems arise, however, in choosing suitable K_d values and demonstrating that their use is actually conservative. Although measurement of a K_d value as defined above seems on the surface to be quite straightforward, the large number of mechanisms involved can lead to many practical difficulties, as will be seen later. As a result, data for relevant rock/water environments are often very limited or poorly defined and, for some important elements, non-existent (McKinley and Hadermann, 1984).

Finally, it should be noted that the nuclear waste literature is full of examples of misuse of sorption data and terminology, particularly of the K_d concept. For example, it is evident that sorption can only occur at any point where rock and water come into contact; that is on the surfaces of fractures and on pore-walls. In many types of fractured rock, the mineralogy of such exposed surfaces is considerably different from that of the bulk of the rock itself. This is because the fissures have always been able to act as conduits for the passage of reacting fluids, and in the geological history of a particular body of rock such fluids may have been hot or concentrated mineralizing liquids. The surface chemistry encountered by the radionuclides may thus be totally different from the bulk rock mineralogy. For example the fissure coatings in granite are often comprised of chlorite, montmorillonite, iron oxides, sulphides and so on, none of which is present in quantity in the primary granite mineralogy. Despite this, many older migration models (and some relatively recent ones) use sorption data (K_d's) measured on crushed host rock, which are obviously inappropriate. The speciation, and hence sorption, of many important radionuclides is also very greatly altered by changing redox conditions. Nevertheless measurements under oxidizing conditions have been used to predict migration in reducing groundwaters.

For the moment, however, it is sufficient to appreciate that on a simple qualitative scale, some elements are sorbed in certain rock water systems much more strongly than others. Plutonium, the *bête noire* of the nuclear industry, is usually strongly sorbed, and in most assessments of migration behaviour (e.g. Hill, 1979; NAGRA, 1985) shows up as being of less radiological significance in eventual releases than radionuclides such as ^{99}Tc or ^{237}Np, which are less well sorbed by most minerals.

While there may be some doubt about our present ability to measure fully realistic sorption values, a qualitative understanding of the behaviour of radionuclide-rock systems is emerging. In some cases, specific groundwater systems at potential repository sites are well characterized, and the sorption behaviour of radionuclides under realistic conditions is known (KBS, 1983). The present generation of computer codes can readily handle quite sophisticated treatments of radionuclide retardation during groundwater transport. The main reasons for the relative simplicity of models used in recent safety assessments (e.g. KBS, 1983; NAGRA, 1985) are the lack of appropriate laboratory data, and the very limited validation of even the simple submodels by *in situ* experiments and natural analogues (see Chapter 11).

THE COMPLETE 'NORMAL-CASE' RELEASE MODEL

After mobiliztion of the waste as a leachate in the near-field, we have followed its progress through the geological barrier in a flowing or diffusing groundwater system. Transport takes place either by diffusion or in response to flow down pressure gradients, and may subsequently lead to release at the surface. During the course of this migration the radionuclides in solution are subject to dispersion and diffusion, and are reversibly sorbed or 'irreversibly fixed' on to the exposed surfaces of the rock. The net effect of these processes is to 'smear out' the concentration-time plot of arrivals at the surface to such an extent that the consequent radiological impact is limited. None of these processes of fixing and retardation described is a truly permanent means of isolating the waste from the biosphere, although radioactive decay means that increasing retardation causes increased destruction of radionuclides within the geosphere, especially the shorter-lived species.

All of these phenomena contribute to the eventual dilution and dispersal of the long-lived components of the waste. These longer lived radionuclides will remain in the repository as a concentrated potential hazard until either this natural dilution has taken place or they have decayed to relative inertness.

The last part of the route: near-surface flow

The far-field barrier with which we have been concerned so far in this chapter has been treated as either a sequence of sediments obeying normal porous medium flow laws, or as a massive fractured body of fairly homogenous rock. In either case the last part of the escape route back to the biosphere may take place through relatively unconsolidated superficial deposits lying on top of the host rocks, and varying very markedly from them in their properties. Weathering products, soils, glacial debris, alluvium and the many other varieties of surface cover can often be several tens of metres thick and show relatively rapid groundwater transport compared with the underlying rocks. Water seeping upwards into a thick sequence of sands and gravels is unlikely to take more than a few tens or hundreds of years to move into surface water bodies, and will mix to some extent with meteoric waters (infiltrating rainfall).

Nevertheless, the importance of this section of the migration route must not be underestimated. Superficial deposits are indeed already used as a host medium for shallow burial of lower activity, short-lived radioactive wastes (see Chapter 9). In the context of deep disposal however, the dilution, dispersion and radionuclide interactions with unconsolidated soils, etc., which occur within this zone, are normally taken into account within the biosphere models. These are considered in more detail in Chapter 10.

MIGRATION FROM A REPOSITORY IN EVAPORITES

Throughout this treatment of migration in the far-field we have concentrated on the way water or mobilized radionuclides move through permeable rocks which obey the Darcy Law of flow, or which allow diffusive movement in an extensive interconnected porosity. The exception to this treatment is the case of evaporite host formations, such as salt domes, which are both impermeable to water and contain very limited free water in pores. In a well chosen unit of thermally stable evaporite minerals with few hydrous phases the only free water would be in the form of an intergranular film, and predominantly as brine inclusions trapped in individual crystals and not interconnected. While these inclusions can move up a temperature gradient, they do not in themselves constitute a radionuclide transport mechanism. In some situations the accumulation of brine around a container of waste may contribute to corrosion, but no flow mechanism is available to transport released radionuclides. Intergranular diffusion in the salt will be exceptionally slow and negligible movement into the far-field is anticipated.

During site selection areas of rapid uplift (diapirism) would be excluded and extensive dissolution of the salt formation can generally be discounted on relevant time scales. Consequently the only natural release scenario is by some form of catastrophic flooding of an operating repository caused by ingress of groundwaters through faults, solution pipes, or access shafts from surrounding rock strata. In such a case the integrity of the salt as a far-field barrier is partially lost and we must consider possible transport through the under or overlying rock, where the processes are identical to those already described. The emphasis here is on 'operating repository', since once the disposal galleries and shafts have been sealed there is no means of access available to waters from overlying aquifer units. Thus, as far as thick evaporite sequences are concerned, provided the seal system is adequate, no sensible migration pathways can be identified which could lead to releases into surrounding groundwaters (with the exception of possible human intrusion into the repository).

EFFECTS OF GEOLOGICAL EVOLUTION ON THE FAR-FIELD BARRIER

To complete this discussion of the far-field geological barrier and radionuclide migration from a deep repository, consideration must be given to the likely effects of climatic and geological change on the rock barrier. It has always been considered important to try to locate sites for repositories such that changes in sea-level, or the geological traumas of an ice age have minimum effect on the performance of the far-field barrier. How significant are evolutionary processes in a safety analysis?

Fundamentally only two scenarios appear to be of any importance to deep disposal. The first is the case of the repository being exhumed and exposed at surface. The second is a radical change in groundwater flow patterns such that throughflow rates are considerably increased and the waste returns to the biosphere more quickly, or with less dilution. Erosion is the only credible mechanism by which the repository could be exhumed on a relevant time scale, short of massive tectonic disturbance, which is itself predictable over very much longer periods (in the order of 10^7 years). Such erosion might accompany very marked changes in climate, leading to very hot or very cold conditions. In either case, the second scenario, that of perturbation of the flow conditions, would occur long before the waste was exposed at surface. The other principal cause of a changed hydrogeological environment would be the occurrence of sustained seismic activity in the region. The reactivation of capable faults (those still potentially subject to movement) may perturb flow patterns but is an unlikely mechanism for causing massive and large scale volumetric flows. Chapter 4 discussed the very limited effects of seismicity on underground structures.

The prediction of the effects of various geological evolution scenarios is based on observations of the effects of similar processes in the past. In the case of glacial effects this is made more straightforward by the fact that we can observe terrain which is presently, or has recently (within the last 10–20,000 years), been glaciated. Such terrain can be in close proximity to, or in very similar geological environments to, areas proposed for disposal purposes. This is especially the case in northern Europe. In fact all of the geological processes likely to be disruptive to a deep repository are active to varying degrees somewhere in the world, and much can be gained from direct study of volcanism, changes in sea and surface levels, major climatic processes, and so on.

The maxim of geology has always been that the present is the key to the past, and that to understand the structure of ancient rock formations we must observe how sediments, volcanic debris, and so on, are deposited today. This can be pushed in the other, albeit unfamiliar, direction of predicting the effects of these processes; what in France are known as 'geoprospective' studies. Very little considered work has been carried out in this area, although there are clear overlaps with probabilistic models such as those described in Chapter 10. Some of the most recent work is brought together by de Marsily and Merriam (1982), in a book of conference proceedings on 'predictive geology'.

Perhaps the main problem lies in the need to be site specific about such predictions, and the parallel need for a substantial database, not only for the site and its surrounding region but also for other occurrences of the phenomena under study. A current study by BRGM in France (Courbouleix et al., 1985) is endeavouring to assemble a database for such exercises, and to develop a methodology for applying predictive geology to specific sites. One of the most convincing attempts to predict future geological changes was that of Vandenberghe et al. (1980), who concentrated on the potential for erosion at the Mol nuclear research site in Belgium as a result of glacial processes which might occur over the next 200,000 years. This study used data on the known depths of erosion in similar sediments in the area as a result of past glacial episodes. This, combined with data on possible rates of subsidence caused by ice loading, isostatic rebound, sea level changes and so on, produced figures on maximum depths of erosion above the proposed repository site.

A further interesting 'case history' concerns the potential uplift of diapiric evaporite structures such as salt domes, discussed earlier in this chapter. Since these structures are in density disequilibrium with their surrounding formations they have a tendency to rise, and as can be seen in parts of the Middle East, to puncture the earth's surface, giving rise to salt hills and salt 'flows'. Under some circumstances, and over very long periods of time, this could clearly be detrimental to any repository contained within them. This problem was thoroughly studied by Gera in the earliest days of the deep disposal programmes (Gera, 1972). Even though many such formations can be observed to be rising at a rate of mm/year, many are presently static, and there is no theoretical reason to believe that rates (i.e. uplift mechanisms) have been markedly different in the past. Several more recent studies have agreed with Gera, that repository behaviour can be reliably predicted provided adequate attention is paid to siting, in particular with respect to:

(a) repository depth relative to present rates of uplift;
(b) uplifts which can be estimated to have taken place in the past;
(c) regional tectonics.

Having gone through exercises such as those described above, it would seem quite possible to site deep repositories in areas where climatic or seismic *effects* can be predicted confidently for the next 10^5 years (although not necessarily the actual climate—it is postulated for example that northern Europe will once again be ice-covered within the next 20–60,000 years). Capable fault zones and areas potentially liable to rapid fluvial or glacial erosion can be avoided. Submergence of a repository land surface below sea level would be a positive advantage since the main driving forces for groundwater flow may cease to operate. Conversely, emergence of the seabed above a submarine repository as a new land surface (as, for example, is anticipated in Sweden) would cause predictable effects on deep groundwater flow.

CHAPTER 7

Field and Laboratory Measurement of the Migration Processes

The concepts discussed in the last chapter must be brought together and tested in order to construct a safety assessment of any potential repository site. This involves measuring many parameters at, or around the site, or in the laboratory, using samples taken from the site. It also inevitably means drilling boreholes to obtain access to the deep geological environment. The general rationale behind site investigation methodology for deep repositories is discussed in IAEA (1982a). In this chapter we examine how such measurements, particularly those concerned with hydrogeological properties, are made. Initially we look briefly at how a basic model of local geological and hydrogeological conditions is constructed, and how this is used to help define a site or regional investigation programme. We then go on to look in detail at how the necessary data are abstracted from such a programme.

The range of techniques mentioned in the chapter is vast, and each has its complexities and limitations of application. Consequently we must be forgiven for not going into detail, and for inevitable omissions in some areas. The aim here is to give a taste of the scope of the science now being used both in the field and in the laboratory. Some useful compilation reports exist. For example, for a thorough and up-to-date treatment of potentially useful site investigation techniques, see IAEA (1985).

REGIONAL FLOW DATA AND THE SITE MODEL

The first essential in assessing groundwater movements in any area is to define the major driving forces for flow, and the possible boundaries to the flow systems. A simple model of the inferred regional fluxes can then be constructed, derived from an analysis of the geology and structure; the topography and climate; known areas of groundwater recharge and discharge; available or inferred values of rock physical properties, and data on hydraulic heads and water chemistry obtained from any wells or boreholes present in the region.

In a well-mapped area of layered sedimentary rocks a considerable amount of the preliminary or 'reconnaissance' modelling which defines potential flow regimes can be carried out by analysis of existing data, with some follow-up geophysical work to aid in the definition of major discontinuities. Clearly, if the site looks promising, this work must be tested in the field by exploratory drilling in areas which the modelling suggest to be complicated in behaviour, sensitive to the values of parameters used in the model, or capable of giving very direct confirmation of some particular result of the model.

This approach can also be adopted in massive crystalline rock areas, but the scope and quality of easily available data are both generally much poorer, and it will often be found that only the major regional faults and other discontinuities have been mapped out. In order to begin to build a model of flow paths in such environments it is essential to carry out fairly detailed ground surveys: mapping and geophysics, to locate the potential flow boundaries. The principal geophysical techniques used in such hard-rock terrain are magnetic, gravity and very low-frequency (VLF) surveys. The first two methods measure anomalies in the earth's magnetic or gravitational field caused by the local inhomogeneity of the rocks. When the data are carefully reduced and plotted on contour maps they provide an 'X-ray' image of the earth's crust to varying depths which can be interpreted to provide data on the presence of faults and other geological boundaries at depth. The VLF technique measures the relative attenuation by different rock types and structures of man-made very low frequency electromagnetic transmissions. These transmissions originate from a limited number of powerful sources which utilise very long wavelengths for communication with submarines. The technique has proved particularly promising for detecting faults in crystalline rocks.

Reflection and refraction seismic profiling are particularly useful techniques for locating faults in sedimentary sequences. They can be used both for shallow surveys, for a near-surface LLW disposal site for example, and for very deep studies (many kilometres if necessary). Clearly they work better if a good seismic velocity contrast exists across a discontinuity in the rock. For detecting faults in clays this is therefore a problem, and sophisticated between-borehole techniques are being developed.

Airborne photographic and geophysical methods provide a 'broad-brush' coverage of large regions and these can be used as a preliminary basis for defining flow regimes. A technique which is presently in the experimental stage for the study of crystalline rocks is the airborne infrared linescan temperature-mapping system (Brereton and Hall, 1983) which enables high-resolution, colour-coded maps to be produced which define small variations in surface-water temperature. These, provided with some measure of surface control, can be used to locate zones of groundwater discharge, some of which may occur along the boundaries to flow cells. A variety of airborne radar techniques have been suggested, but not fully proven, for 'stripping' vegetation and surface cover to provide either a radar image of bedrock (and its major fractures) or a radar image of the water table. Such techniques still need to be evaluated fully.

It is clearly not practical to go to the considerable effort and expense of making an accurate appraisal of all the suspected groundwater flow paths in a particular study area unless it is eventually decided to attempt to prove it as a repository site. In that case it can be seen that the area of consideration in a sedimentary sequence will need to be much larger than in crystalline rock terrain, simply because the path lengths are longer.

In carrying out generic research (as distinct from site investigations) it is more useful to make detailed studies of particular concepts of the models discussed above. The models of groundwater flow need to be tested by exploratory drilling to see, for example, whether faults do have very widely varying permeabilities and whether these can act as flow boundaries and under which conditions. Only when a wide cross-section of hydrogeological environments have been tested will it be possible to develop the comprehensive flow model to the level of confidence required for performance assessment.

HYDROGEOLOGICAL PROPERTIES OF THE GEOLOGICAL BARRIER

Hydrogeological measurements are largely carried out in boreholes which are designed to penetrate different regions of a predicted groundwater environment. Such boreholes are the crux of the whole assessment of the far-field barrier, as they enable both sampling of the rock and groundwater at depth, and *in situ* measurement of all the physical and chemical properties of the rock-water system required for performance modelling. Considering their importance, it is worthwhile taking a closer look at how they are made and what they can do.

Drilling boreholes

While a considerable amount of information about a geological environment can be obtained by surface examination and remote geophysical techniques, it is clearly essential to sample and make meaurements in three dimensions rather than two, if that environment is to be fully understood. Despite the fact that a 10 cm-diameter borehole 1000 m deep provides such apparently tenuous access to the depths (it can be compared in scale to a human hair more than a metre long) it affords remarkable scope for obtaining very detailed information about the rock, using sophisticated 'down-hole' instruments, which can examine some properties of the rock almost as well as if it were in the laboratory.

In many hard rocks boreholes are self-supporting; in other words they can be drilled and left 'open' to considerable depths provided there are no strong regional tectonic stresses. Borehole walls can thus be examined directly by a 'logging tool'. In mixed sedimentary sequences where the strata have different mechanical properties, an open borehole would quickly deform, or simply close up by flow of the surrounding rocks. Deep holes in such environments have to be supported as they are drilled and then lined by insertion of a metal pipe (called casing) throughout the depth of the hole (Fig. 7.1). Prior to casing a hole it is supported by drilling mud, a dense slurry which not only lubricates the bit during drilling and carries the rock chippings back to the surface as it is pumped and circulated, but also holds the borehole walls apart. Clearly, any logging tool which requires direct contact with the rock must be used prior to casing, while the drilling mud is still in the hole.

Examining uncased boreholes in crystalline rocks is obviously a simpler task. Since the water table may be almost coincident with the surface all equipment must function underwater. Boreholes in these rocks can be drilled in various ways, the principal techniques being those which 'core' the hole (obtain a stick of rock almost the same diameter as the hole, see Fig. 7.2), and those which simply

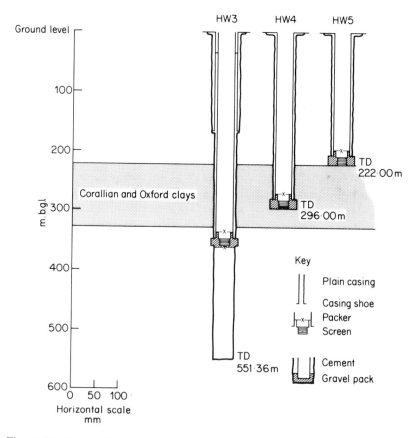

Figure 7.1 Schematic representation of experimental borehole completions in sedimentary rocks, taking as an example those used to study groundwater movement and properties beneath the Harwell site in England (from Robins et al., 1981). Perforated screen (see Figure 7.5) was placed in these boreholes to extract water from specific depth intervals, in and around the clay formation of interest

Figure 7.2 Geologists examining rock core obtained during the course of drilling in hard crystalline rock

grind all the rock away, producing only dust and chippings (a technique known as 'rock-bitting'). Cored boreholes obviously provide the maximum information as data from remote tools examining the borehole walls can be compared directly with the contiguous piece of rock core which has been brought to the surface. However, cored holes can be a factor of ten times more expensive to drill, and progress is slower and more difficult. Downhole 'logging' tools and methods are therefore developed to fit the exact requirements of specific parameter measurement. This may also allow us to evaluate how little cored hole is really necessary to obtain the data required. When drilling a cored or rock-bitted hole in crystalline rock the cuttings must be extracted and the bit kept cool. This 'flushing' of the drilling face is usually done by either compressed air or water. There are limitations on the use of both methods, and variations in technique, but they are mentioned here because the effect of introducing either air or surface water to the deep environment must be taken into account when making any subsequent hydraulic or chemical measurements at depth. Generally a period of recovery is required, during which air pockets or drilling water are naturally flushed from the groundwater system, before any serious measurements can be attempted. In the case of water chemistry this may take many months.

Because a borehole in crystalline rock often penetrates a fairly uniform hydrogeological environment, it is not generally necessary during drilling to isolate any one section of the hole from any other. An obvious exception to this is the soil and weathered rock cover (perhaps the upper few tens of metres or less) through which considerable near-surface flow can take place. To avoid contamination of the deeper environment by surface waters the upper part of the hole would be cased off. In the mixed sedimentary environment it is advisable to isolate the various units of interest if certain measurements are intended to be made, and to ensure that water mixing and contamination up and down the borehole does not take place. The complete casing of the hole described earlier is not sufficient to ensure this, as flow can take place behind the casing, so cement is often pumped into the annulus between casing and borehole wall to seal off systems, although this may considerably perturb surrounding water chemistry. As drilling mud is frequently used as the flushing medium when drilling either cored or rock-bitted holes in such rocks, this must be removed and washed out as casing proceeds, so that it does not contaminate the borehole geochemistry. Often this flushing is monitored by adding an easily measured tracer to the drilling mud.

Hydrogeological measurements

Hydraulic testing:

The groundwater head and rock hydraulic conductivity (K) are usually required for particular depth intervals in a borehole. Bulk hydraulic conductivity for a complete borehole can be measured, but as this is an average for a large volume of rock it can only be of limited value. In fractured rocks, it is important to consider particular fractures or fracture sets and measure head and conductivity in each, which combined with data on fracture surface chemistry and water chemistry can be used directly in the flow model. It is also useful to measure the Darcy parameters for zones of intact (unfractured) rock, and to measure them as a function of depth, or measure their relation to particular discontinuities in the rock such as fault zones, dykes or other changes in rock type. In sedimentary rocks the same parameters must be measured for specific formations or specific horizons within a particular formation, and anistropy must also be considered (e.g. differences in horizontal and vertical hydraulic conductivity). To do this sections of the hole must be isolated, and this is done by using pairs of inflatable (or mechanical) 'packers' held together by a rigid rod (Fig. 7.3). A packer system can be moved up and down a borehole when deflated, but seals against the wall when air, or more often water, is pumped into the packer elements. The region between the packers is thus isolated and can equilibrate in terms of flow properties and pressure with the rock in that zone. Transducers placed between the packers and above and below them measure the local groundwater hydraulic pressure. It is often found that particular fracture systems are at different pressures from those which might be estimated simply by virtue of their depth (i.e. depth multiplied by water density) and such measurements can be used to build up a picture of local hydraulic gradients and degrees of interconnectivity of fractures. Various methods are used to measure K, perhaps the best as regards minimizing perturbations in the borehole environment being an extraction

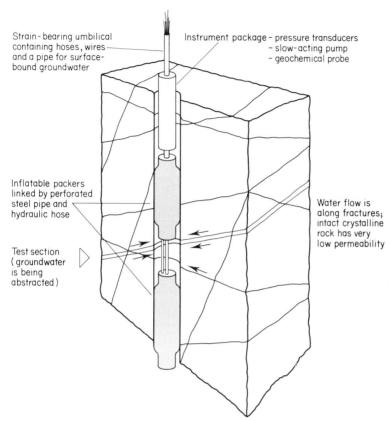

Strain-bearing umbilical containing hoses, wires and a pipe for surface-bound groundwater

Instrument package – pressure transducers
– slow-acting pump
– geochemical probe

Inflatable packers linked by perforated steel pipe and hydraulic hose

Water flow is along fractures; intact crystalline rock has very low permeability

Test section (groundwater is being abstracted)

Figure 7.3 The principle of measuring groundwater parameters in a borehole in fractured rock using inflatable packers to isolate a zone (in this case a specific fracture) of interest (after Black and Chapman, 1981)

technique (e.g. Holmes, 1981), whereby water is pumped out of the packer interval and the local pressure variations and extraction rate are monitored. Alternatively, water can be injected under a known overpressure, and K calculated from observed flow rates. These basic techniques are well established in hydrogeology, but their application in low-hydraulic conductivity environments is quite novel. Consequently the technology has required upgrading in terms of the precision and working limits of the equipment. Measurements of K are now being made routinely in rocks which have previously been regarded as 'impermeable' in that K is so low that the rocks were of no interest as aquifers; in fact so poorly permeable as to have been of no hydrogeological interest whatsoever.

Borehole logging:

Hydraulic tests are generally made every few metres in a regular fashion as the packer unit is moved along the hole. In order to relate this to the known physical properties of the rock, individual test sections must be correlated with either a fracture log of the core, with the core itself, or with some physical record of the nature of the borehole wall. As core is extracted from the borehole it is examined in detail and 'logged'. A core log consists of a description of the texture, structure and mineralogy of the rock, a record of the number, orientation and degree of 'openess' of its fractures, and other details of its properties. Provided the core can be well oriented with respect to its position in the hole, then one has a reliable measure of the fractures being tested hydraulically. However, this is rarely practicable, indeed in zones of badly fractured rock it is often impossible to reconstruct the core at all. Consequently a number of logging tools are run in the borehole to act as controls on the core log by examining the nature of the borehole walls.

The simplest is the caliper log which measure the width and roundness of the hole by means of a multiple armed, spring-loaded caliper. It can also pick up voids due to open fractures or broken rock. Down-hole television is a direct means of observing the hole. Miniature colour or monochrome cameras with their own light source can be used to examine the borehole wall. The most sophisticated method available at present is the acoustic televiewer which produces what is effectively a sonar image of the complete borehole wall, 'unwrapped' such that a fracture cutting the hole at an angle appears as a sine-wave on the print-out (Fig. 7.4). This method records openness, spacing and orientation of every fracture down to those only about 10 microns wide. When these logs are run in a borehole and combined with standard logs of borehole verticality and direction (boreholes are rarely straight) then a detailed picture can be built up of the local rock structure. This is essential, not only to enable reliable placement of packers against undamaged rock, but also to interpret the results of the hydraulic packer testing.

A further Darcyan parameter requiring measurement is the bulk porosity of the rock: that space occupied by water in pores and open fissures. The neutron-sorbing properties of water are used as the basis for the neutron porosity log, which is essentially a sealed radioactive source which emits neutrons whose sorption and scattering by the rock is measured in the borehole. This gives a direct correlation with water content and hence porosity, but needs careful calibration with the rock chemistry, since water is not the only molecule to sorb or scatter neutrons. At present this technique works well in certain sedimentary rocks but is not fully calibrated for crystalline rocks. Consequently a laboratory method is used to determine intact rock porosity, but this takes no account of the smaller fissure porosity. The rock sample from the core is dried and its pores evacuated of water and air. It is then resaturated with water or mercury and reweighed to give a direct measure of total porosity. It should be noted that the neutron log measures total porosity while core-resaturation measures connected porosity which, as previously discussed, may not be quite the same thing.

The temperature-conductivity log is simply a continuous record of the temperature and conductivity of the water in a borehole as a function of depth. Naturally it can only be of real use in an uncased hole and is best employed when

Sonic televiewer (SABIS)
Fissures and fissure
systems numbered

Azim. 270   360   90

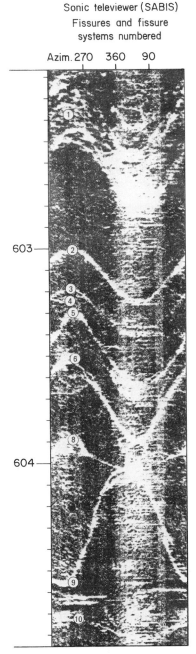

Figure 7.4 Typical output of the borehole televiewer; a sonic picture of a borehole wall in granite, 'unwrapped' so that fractures cutting the hole at an angle appear on the trace as sinusoidal curves. On this trace no significant zones of broken rock or void spaces can be seen. The figures at the side indicate borehole depth in metres. *Reproduced by permission of Nagra, Switzerland*

conditions have been allowed to equilibrate. Essentially it indicates zones of water inflow and mixing in a hole, and shows up zones of exceptionally low K and hence low flow, where conditions are very stable.

When carrying out the hydrogeological testing of a cased hole in a mixed sedimentary sequence a certain number of the logs described above must be run before casing takes place. All the packer testing can, however, be carried out after casing and cementing is complete and the hole has been established. This is made possible by the use of screened casing which is perforated to allow passage of water (Fig. 7.5) and which can be emplaced in zones or formations of interest. In addition normal casing can be perforated at any point by the use of miniature, shaped, explosive charges which provide holes of known diameter and which penetrate both casing and the cement behind. Casing and screening a deep borehole is a complicated process which requires a great deal of planning and forethought on the part of the hydrogeologist.

Cross-hole testing and tracers:

A further important technique is provided by 'cross-hole' geophysical and hydraulic testing; measuring properties between pairs or groups of closely adjacent boreholes. The principal use of hydraulic techniques in such configurations is to measure permeability anisotropy, and to examine large volumes of intact rock which have not been perturbed by the drilling of holes (Black and Barker, 1981b). The injection and monitoring of tracers and their movement is also made possible by such borehole configurations, and it is general practice to characterize the experimental rock volume geophysically and hydraulically prior to injecting tracers. Such techniques have been a central feature of the testing programme in the Stripa underground laboratory in Sweden and are discussed again later in this chapter.

The use of tracers injected into groundwaters is one of the most direct methods of obtaining flow parameters. Many tracers are available, including dyes, brines, short-lived radionuclides and large, chemically inert molecules. Combinations of tracers which do not interact with the rock (so-called 'conservative' tracers) with sorbing species, can provide information on the likely behaviour of leached radionuclides in the ground. The interpretation of such experimental data, usually derived from inter-borehole tests, is difficult and the results are often ambiguous, in that similar concentration vs time profiles of tracer 'breakthrough' at a monitoring point can be produced in different ways as a result of the interaction of the various retardation mechanisms. However, the results of the best-controlled tracer tests in simple formations can be interpreted with reasonable confidence (e.g. Williams *et al.*, 1985). Some results must, however, be regarded with caution, as the more complicated variants (with concentration-dependent sorption, matrix diffusion and multiple migration paths) often have so many free parameters that their interpretation can be an exercise in curve fitting, which cannot be extrapolated outwith the experimental conditions involved.

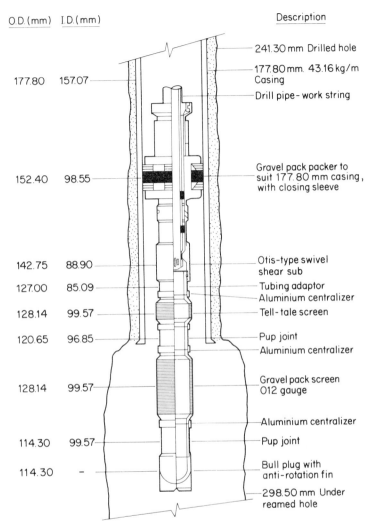

O.D.(mm)	I.D.(mm)	Description
		241.30 mm Drilled hole
177.80	157.07	177.80 mm. 43.16 kg/m Casing
		Drill pipe- work string
152.40	98.55	Gravel pack packer to suit 177.80 mm casing, with closing sleeve
142.75	88.90	Otis-type swivel shear sub
127.00	85.09	Tubing adaptor / Aluminium centralizer
128.14	99.57	Tell-tale screen
120.65	96.85	Pup joint / Aluminium centralizer
128.14	99.57	Gravel pack screen 012 gauge
		Aluminium centralizer
114.30	99.57	Pup joint
114.30	–	Bull plug with anti-rotation fin
		298.50 mm Under reamed hole

Figure 7.5 Schematic diagram of the complex completion which can be used at the bottom of a cased borehole in sediments to allow sampling of groundwaters from a particular formation. A perforated casing 'screen' is installed in an oversized section of hole which has been flushed and filled with clean sand or fine gravel to filter influent water (from Robins, *et al.*, 1981)

MEASURING PARAMETERS INVOLVED IN DISPERSION AND MATRIX DIFFUSION

In flow modelling only the *kinematic* porosity is required, and techniques for measuring this must be related to hydraulic testing in boreholes as discussed above. The *total connected* porosity of a sample of rock is of prime concern from the viewpoint of diffusion, and can be measured in the laboratory by liquid-invasion methods such as those mentioned earlier.

Laboratory measurements of solute diffusivities can be made on small rock samples (Skagius and Neretnieks, 1986) and, in some cases, related to their porosities and pore structure (constrictivity and tortuosity). Such tests can be carried out under confining pressure to simulate *in situ* behaviour more accurately. In granitic rocks this appears to have little effect on the measured diffusivities (Bradbury and Stephen, 1986). Treatment of samples can, however, cause problems, because an increase in porosity may be caused by the sampling processes, or by stress relief which is not subsequently removed by repressurization of the sample. For many rock types confirmation of laboratory measurements by *in situ* experiments is required, but the latter are complicated by the long time scale of diffusive processes. Laboratory experiments generally measure the build up of an elemental concentration profile by diffusion from a spiked solution into a rock sample. This is affected by various retardation processes, and a value called the apparent diffusivity is measured. A few experiments measure the 'steady-state' flux of solute resulting from a set concentration profile, from which a pore diffusivity (D_p) is derived which is not affected by retardation. The pore diffusivity, D_p, is related to the apparent diffusivity D_a by a retardation factor. In cases where the K_d concept discussed previously can be assumed to hold,

$$D_a = D_p/(1 + K_d \, \rho \, (1 - \varepsilon)/\varepsilon)$$

where ε is the porosity and ρ the density of the rock. The geometry of the specimen must be accounted for when calculating actual diffusion rates and a sophisticated mathematical treatment is necessary in most situations (see for example, Neretnieks, 1982). Diffusivities of cations and anions from the dissolved waste, together with those of colloids, can be measured. The latter are generally negligible for all but very small particles (molecular weights of a few thousand daltons) except in very open-pored rocks.

It would obviously be preferable to be able to test the diffusive delay model in the field and various attempts have been made to do just this. Unfortunately, the coupling of this phenomenon with sorptive retardation of the tracer used makes interpretation difficult. In such experiments both sorbing and non-sorbing tracers are released in a controlled manner along a characterized fissure and their arrival at a separate point in the fissure monitored. The migration route can also be exhumed after the experiment by drilling (Abelin, *et al.*, 1986), and the extent of migration of the tracer out into the intact rock bounding the fissure measured directly.

When deriving laboratory values of these parameters, or trying to interpret field tests, it is useful to have some idea of the nature of the porosity being studied, as well as its simple numerical value. Important factors include interconnectivity, tortuosity, pore sizes and pore wall coatings (chemistry and mineralogy). It is also possible to measure the variations in porosity around specific fractures or within a zone of intact rock, using core material. Since different values of diffusivity will be found in a variable-porosity rock, this factor is important in a thorough analysis of, for example, a field tracer test.

Dispersion is largely controlled by the fracture pattern of the rock or by its pore structure and distribution, and includes both hydrodynamic and physico-chemical processes. The dispersion coefficient is a scale-dependent parameter which has to be deduced from *in situ* measurements. Clearly, in any given volume of rock, the value determined will vary with the scale over which the experiment is performed, and for far-field transport calculations those determinations made on a large scale will be the most appropriate. Such measurements are normally performed in pairs or groups of boreholes, and are sufficiently complex and time-consuming to be warranted only at the detailed site investigation stage. Single borehole, small-scale, rapid reconnaissance methods are also available, and a technique is described by Noy and Holmes (1986), who also review some of the many inter-borehole tracer techniques for dispersion measurement developed over recent years.

In order to interpret dispersion data in fractured rocks, it is useful to be able to model the fracture pattern. Obtaining this information means recourse to a combination of surface-fracture mapping and borehole-fracture logging. Some of the latter techniques have already been discussed: fracture logging of core, borehole TV, acoustic televiewer and so on. In addition, the borehole wall measurements can be extended into the surrounding rock using 3-D sonic logs and borehole-to-borehole sonic and radar techniques which can define the orientation and spacing of fractures at distance.

DETERMINING HOST-ROCK AND GROUNDWATER CHEMISTRY

The geochemistry of the host rock provides a range of essential information for site evaluation purposes:

(1) The bulk groundwater chemistry influences the performance of the near-field barriers, radionuclide speciation and solubility limits, and in some cases can be used to place some constraint on the nature and evolution of the hydrogeological environment. Trace species in groundwater may give much stronger indications of its origin and history, which can constrain the hydrogeology, and in some cases may even allow the dating of waters from some part of the groundwater regime.

(2) The geochemistry of the rock itself may indicate the extent to which it has been in contact with groundwaters over very long periods, and might consequently be in contact with any transported radionuclides in the future. It is also important to understand the role of the rock in buffering water chemistry and chemically retarding the migration of radionuclides.

The procedures for sampling deep-rock and groundwater systems vary somewhat, depending on the host rock and the geological environment. This can be illustrated for the two quite different cases of fractured rocks and plastic clays.

Most deep samples are obtained from boreholes. Rock samples are taken from the drill core, while flowing groundwater can be sampled using the packer systems described earlier. Sampling cartridges or pumps can be installed in these systems to obtain water from particular zones or fracture systems, or units of different lithology.

In crystalline rocks most water transport occurs in distinct fissures which are thus of fundamental interest, both from the viewpoint of the water in them and because of their surface chemistry. During coring, however, such fissures represent zones of weakness at which the core may break, losing important surface material. Even if the core does not break, drilling fluid may penetrate the fissure and alter its surface properties. Although the net flow of water through crystalline rocks of interest for disposal is generally very low, its extremely localized nature makes sampling somewhat easier. Nevertheless, several problems exist:

(1) Drilling fluid may penetrate the fissure, and for low flow systems recovery of the original water chemistry may take many months. Tracers (e.g. Li, or Uranine dye) may be added to the drilling fluid to identify such contamination.

(2) Groundwaters may be very poorly chemically buffered, and important chemical parameters (especially pH or redox conditions) may be altered by solid contaminants, casing or linings in the borehole.

(3) If water is allowed to depressurize during sampling, its chemistry may change considerably. The pressure relief causes degassing, which may even be visible as effervescence. The loss of carbon dioxide, for example, can cause a significant rise in pH.

One way round the problem of depressurization and the general perturbation of delicate *in situ* geochemical equilibria is the use of downhole probes, which can measure several important parameters, such as pH, redox conditions, temperature, conductivity and the partial pressures of some gases. The operation and calibration of such *in situ* monitors still presents many problems, however, and their use is by no means routine.

In a plastic clay formation, where advective water fluxes may be negligible, both water and rock samples may have to be taken from the drill core. Penetration of drilling fluids into the core can also be problematic when using this material, and some trimming of the margins is usually needed before samples can be stored. Extraction of the pore-water for separate characterization is, however, the main difficulty, as present techniques (such as squeezing, centrifuging or displacement by an immiscible fluid) will tend to disrupt the complex equilibria in the pore systems (Brightman *et al.*, 1985). Additional interpretational problems arise because much of the water in such clays may be close to charged mineral surfaces, and behave differently from 'free' pore waters.

Assuming that a good water sample can be obtained, the first requirement is for an analysis of the major components in solution (e.g. $Na+$, $K+$, Ca^{2+}, Cl^-,

SO_4^{2-}, CO_3^{2-}, etc.), the pH and redox conditions. In addition, a range of trace species can also be measured:

(1) Radionuclides which are derived from the atmosphere (e.g. ^{14}C and ^3H) and hence give some indication of the 'age' of the groundwater. While the presence of short-lived radionuclides such as ^3H or anthropogenic chemicals such as fluorocarbons indicates the presence of relatively 'young' waters, mixing of waters of different origins and ages generally makes exact dating impossible.

(2) Radionuclides derived from the rock (members of the ^{238}U, ^{235}U, and ^{232}Th decay series) can also give some indication of water age and also the prevalent chemical environment. For example, U is much more soluble under oxidizing conditions and hence may be used to confirm the results of other measurements of redox conditions.

(3) The ratio of stable isotopes of some light elements (e.g. $^2H/^1H$, $^{13}C/^{12}C$, $^{18}O/^{16}O$, etc.) varies between waters from different sources (for example in rain, as a function of temperature at the time of the rainfall), and also as a result of rock/water interactions. This again can be used to evaluate the history of the groundwater.

(4) Trace organic compounds, micro-organisms and colloids. These are possible perturbing factors in the normal-case model, and should be characterized, if possible, in water samples. They are grouped together here because as yet their analysis is very difficult, would have recourse to related techniques, and is the focus of current research.

The analytical requirements for examining rock samples are, in principle, the same for any type of geological material, although the procedures used may be somewhat different in each case. Although standard mineralogical techniques generally present the composition of a particular rock in terms of a bulk, homogenized analysis of the elements present, or as a modal analysis of the minerals present, this is not usually very useful. Much more useful are data on the particular rock surfaces which would come into contact with migrating radionuclides, and in the case of fissure flow, profiles of mineralogical/chemical characteristics away from the fissure, into the bulk rock. Very careful sample selection must be combined with a range of bulk (e.g. X-ray diffraction, X-ray fluorescence, differential thermal analysis, etc.) and surface (e.g. scanning electron microscopy, secondary ion mass spectroscopy, proton-induced X-ray emission spectroscopy, etc.) analytical techniques. In order to aid interpretation, concentrations of particular trace elements may be measured and compared to those in the groundwater. The natural series radionuclides are particularly useful in this respect.

The measurement and interpretation of natural radioisotopes and stable isotope ratios are described in more detail in general textbooks (e.g. Ivanovich and Harmon, 1982; Fritz and Fontes, 1980) as are more conventional water (e.g. Stumm and Morgan, 1981) and rock (e.g. Zussman, 1977) analysis techniques.

MEASUREMENT OF RADIONUCLIDE RETARDATION PARAMETERS

In our previous discussion of radionuclide/rock interaction ('sorption') processes, we noted that a number of mechanisms were involved, and these can be subdivided into uptake processes, which involve transfer of radionuclides from solution to rock, and retardation processes, which slow down migration of solute without any actual uptake occurring. Experimentally, the former may be examined by 'static' methods, while alternative 'dynamic' experiments measure the net result of both processes.

Before continuing to describe the experimental methods, it is important to note that most retardation parameters are very dependent on the chemistry of the rock/water system studied, and hence very careful sample selection and control of parameters such as pH, temperature and redox conditions is required. This is especially important for studies of the actinides (whose aqueous phase speciation varies greatly with quite small changes in pH/redox conditions), and for rock/water systems which are poorly buffered chemically.

Static techniques effectively measure the partitioning of a radionuclide spike between solid and liquid phases. Ideally, exchange of radionuclide between phases should be independent of any water fluxes, and thus pure 'sorption' is measured. A particular form of static technique, batch equilibration, is by far the most popular method for quantification of sorption, and is responsible for the majority of data reported.

The principle of the batch equilibration methodology is extremely simple. A rock sample is added to a radionuclide-spiked solution, and a partition coefficient (often loosely referred to as a distribution coefficient, or 'K_d') is calculated from the ratio of radionuclide concentration in the rock to that in solution.

Although subject to practical limitations, the reaction time of such experiments should be as long as possible to allow the system to come to equilibrium, both before and after addition of the spike, or at least to ensure that the rate of any subsequent reaction is negligibly slow. For finely divided rock samples the time required is generally in the order of days to weeks, but for monolithic samples of rock reaction may continue at a measurable rate for many months or years, owing to slow diffusion processes into the rock matrix. During the course of such experiments, it is inevitable that slow hydrolysis reactions take place between rock and water, and the reaction products may also be involved in radionuclide sorption. At low temperatures the rates of such reactions are extremely slow for most rock types, although they may need to be taken into account in interpreting results in some cases.

Although reversibility of sorption is generally assumed, experimental studies indicate a spectrum from complete reversibility to total irreversibility in some systems. This can generally be examined quite simply by either diluting or replacing the spiked water after a sorption experiment and measuring the radionuclide partitioning after further equilibration.

The results of batch sorption experiments are often reported simply as 'Kd' values, but the data produced from this technique can be more usefully expressed as sorption functions, relating the measured partitioning to a range of variables. When sorption is expressed as a function of radionuclide concentrations at a constant temperature, these functions are called sorption 'isotherms', and are increasingly reported.

In practice, large numbers (hundreds or thousands) of batch sorption/desorption experiments are often run in parallel, to assess variations with several important parameters, such as temperature, time, pH, concentration, competing species and so on. As any geological system will show evolution in geochemical conditions over the time scales being considered (particularly in the near-field), the only practical experimental technique capable of examining the magnitude of change of sorption for all likely combinations of perturbing parameters is some form of batch methodology.

Other forms of static experiment have also been devised for specific applications. One of the most useful is surface autoradiography (e.g. Vandergraaf *et al.*, 1982) in which a rock surface is exposed to a spiked solution and the spatial distribution of sorbed activity is detected by a special radiation-sensitive film which is contacted with the surface after equilibration (Fig. 7.6). This distribution can be related to the mineralogy and pore structure of the rock and the parameters controlling sorption better interpreted. It is particularly useful if mineral phases with very low abundances in the rock are responsible for sorption reactions.

The basic 'dynamic' technique is the column experiment, in which a radionuclide-spiked solution is passed through a rock or soil column and either the subsequent radionuclide concentration profile (as a function of time) in the effluent solution, or the profile with distance down the column, is used to determine a retardation factor (Fig. 7.7). For naturally disaggregated materials in which porous flow occurs (e.g. sands), columns can be prepared with minimum physical perturbation directly from cores, although anisotropy (e.g. bedding structures) must be considered. For solid rock cores there are problems of preferential flow down the sides of the specimen and alternative techniques involving crushing the rock and packing it into the column are often used. The relevance of such studies to real systems is very limited or even negligible. For low permeability rocks, high hydraulic heads are often used to increase flow within a column, but there are again problems in relating this to reality since flow paths may be forced open by the pressures applied.

Alternative techniques involve essentially static equilibration, followed by evaluation of diffusion into the rock matrix (e.g. Torstenfelt *et al*, 1982), or diffusion cells which measure transport through thin sections of rock (e.g. Bradbury *et al.*, 1982). In both these techniques migration distances are generally so small that there can be problems in ensuring that pathways are not artefacts of sample preparation procedure. In fissured rocks, columns may be designed to allow flow through either a natural or an artificial fracture. Practical problems involve excavating and transporting a fractured block of rock to the laboratory

150

and ensuring that it can be reassembled under realistic confining pressure without irreversible damage having occurred to the fracture surface or materials.

Details of particular column methodologies vary widely, but use of a single term (in this case 'retardation factor', or 'R') for a range of totally different parameters can create some confusion. A retardation factor can be described in a strict sense only when reversible, concentration-independent sorption is the sole retarding mechanism. The assumption of such a mechanism allows a K_d value to be derived from the retardation factor, R, and experimental results are often reported in this way.

(1)

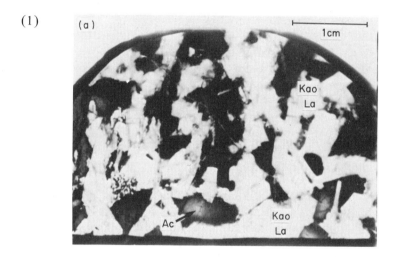

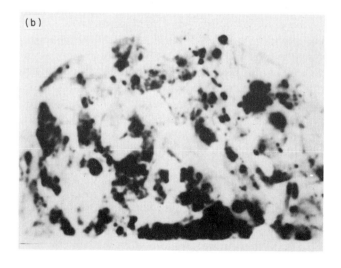

(2)

Figure 7.6 Examples of autoradiographs of radionuclides sorbed on to cut surfaces of rock. (1) Photomicrograph (a) and autordiograph (b) of ^{137}Cs sorbed on to a thin slice of gabbro from the East Bull Lake pluton, N. Ontario, Canada. Dark areas of the autoradiograph indicate where sorption of Cs has occurred. Ac = actinolite, kao = kaolinite, La = laumontite. (2) A similar pair (b is the autoradiograph) showing sorption of ^{147}Pm on to a thin slice of highly fractured granite from the Eye-Dashwa lakes pluton, near Atikokan, N. Ontario, Canada, with dark zones indicating where sorption has occurred preferentially. Qtz = quartz, Ti = titanite, Ep = epidote (courtesy of Atomic Energy of Canada Ltd)

152

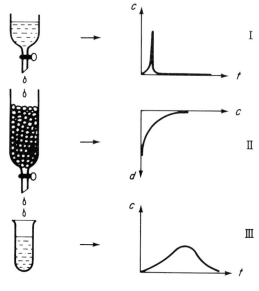

Figure 7.7 Highly schematic depiction of a column sorption experiment. Groundwater containing a radionuclide 'spike' is injected into a packed column or core sample of rock or soil. It may enter the column as a pulse when plotted against time (I). On passing through the column its concentration in solution decreases with distance travelled as sorption occurs (II) and the effluent solution displays a spread of concentration with time (III) owing to dispersion and retardation in the rock. This is a similar pattern to that observed in tracer experiments in the ground. *Reproduced by permission of Nagra, Switzerland*

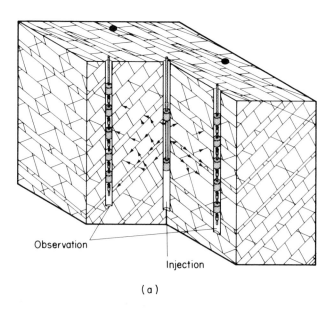

(a)

Figure 7.8 (a) Simplified diagram of an *in situ* tracer test in fractured rock, with a tracer solution injected between packers from a fixed interval in a central borehole, and monitoring points within various isolated zones of surrounding holes. The distribution of flow within different parts of the fracture system can be assessed. *Reproduced by permission of Nagra, Switzerland.* (b) An equivalent *in*

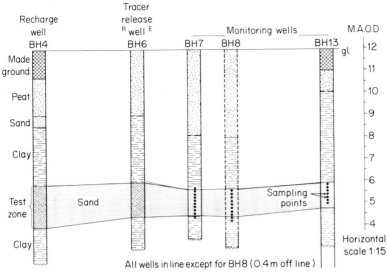

Details of field experimental array

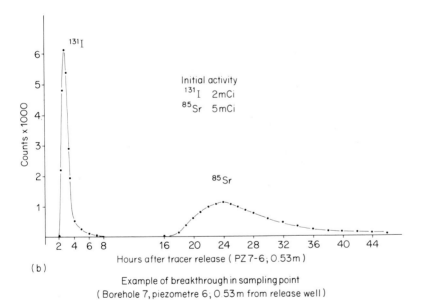

(b)

Example of breakthrough in sampling point
(Borehole 7, piezometre 6, 0.53 m from release well)

situ radionuclide migration experiment in a shallow sand aquifer (Williams *et al.*, 1985). Non-sorbing ^{131}I and sorbing [85]Sr are released from a borehole and carried 'downstream' in the sand aquifer unit by water injected into an adjacent borehole (recharge well). The curves show the arrival concentrations and times at a downstream monitoring point. The conservative I tracer arrives first as a pulse, while the sorbing Sr displays an extended curve, with lower concentrations, as a result of retardation during transport (courtesy of British Geological Survey)

We have seen that the mechanistic significance of radionuclide partition coefficients is very dependent on the design of the experiment that measures it, and a wide range of different parameters are commonly reported as 'K_d'. Their use is further complicated by the wide scatter of measurements, even in inter-laboratory comparisons where techniques, conditions and materials are supposed to be standardized. One commonly quoted, though by now rather old example (Relyea and Serne, 1979), produced a range of measured K_d values for a single rock-water-radionuclide system, spanning five orders of magnitude. Nevertheless, such batch techniques are becoming more reliable and sophisticated. A database of measurements (with extensive supporting background data) has been developed by the OECD/NEA (ISIRS—International Sorption Information Retrieval System) which allows a number of different sorption isotherms to be extracted.

Many other variants of these experimental techniques exist, and have been reviewed by McKinley and Haderman (1984). As different methods measure slightly different processes, and are appropriate to different rock types, a range of complementary techniques must be chosen carefully, to characterize sorption or retardation in any particular environment. Once a simple system is quantified as a function of the important environmental variables, it can be compared to the behaviour of more complex, multicomponent systems. Eventually it should be possible to quantify the behaviour of actual waste leachates as they move through the specific geochemical environment of a repository site. Nevertheless, laboratory experiments are inherently limited in scale and in the accuracy with which they can simulate the real environment, and complementary field experiments are required.

Field migration tests have been devised to try to simulate the transport properties of relevant radionuclide species, principally using inter-borehole injection and monitoring techniques which measure movement of the input spiked solution, either through the body of the rock as a whole or through a characterized fracture, or fractures (Fig. 7.8). Such tests have been run in a variety of rock types, usually using short-lived isotopes (Côme et al., 1985). Tracer tests of this type have been mentioned several times already in this chapter, each time as a means of measuring a different rock or flow parameter. Because one is using a real rock-water system, the test interpretation must take into account all the other factors which govern transport apart from the radionuclide/water interactions: porosity, flow rates, diffusion, dispersion, decay, pathlength and so on.

Owing to the many complexities discussed above, there are shortcomings at present to all the experimental techniques associated with the behaviour of leached radionuclides in contact with rock, and this is an area where there is likely to be considerable difficulty in providing tight limits on parameter input for the release and migration model.

UNDERGROUND RESEARCH LABORATORIES

While many of the factors discussed above can best be studied in the laboratory or in deep boreholes, there are many advantages in being able to carry out experiments deep into the rock. They can then be performed under similar physico-chemical conditions to those expected in a repository and make use of large volumes of intact rock. Underground research laboratories (URL) are thus constructed to give access to the deep geological environment and are now operating (or planned) in many countries, in many rock types.

URL have been in operation since at least the mid-1960s. Major facilities which have been, or are being, used (or are in advanced stages of construction) are listed in Table 7.1, and examples are illustrated in Figs. 7.9–7.11. It is likely that this list will expand considerably in the next few years, when assessments of potential repository sites which involve URL construction are initiated in several countries (e.g. USA and Switzerland).

Table 7.1   Major past or present underground research laboratories

Rock formation	Laboratory name	Country
Salt		
(bedded)	Salt Vault (Kansas)	USA
(dome)	Avery Island (Louisiana)	USA
(dome)	Asse	FR Germany
(bedded)	WIPP (New Mexico)	USA
(dome)	Hope	FR Germany
Crystalline rock		
(granite)	Stripa	Sweden
(granite)	Grimsel	Switzerland
(granite)	Edgar Mine (Colorado)	USA
(granite)	URL (Manitoba)	Canada
(granite)	Climax Mine (Nevada)	USA
(granite)	Fanay Augeres	France
(granite)	Akenobe Mine	Japan
(basalt)	NSTF (Washington)	USA
(tuff)	G-tunnel (Nevada)	USA
Argillaceous rock		
(plastic clay)	Mol	Belgium
(clay-marl)	Pasquasia	Italy
(mixed sediments)	Konrad Mine	FR Germany

156

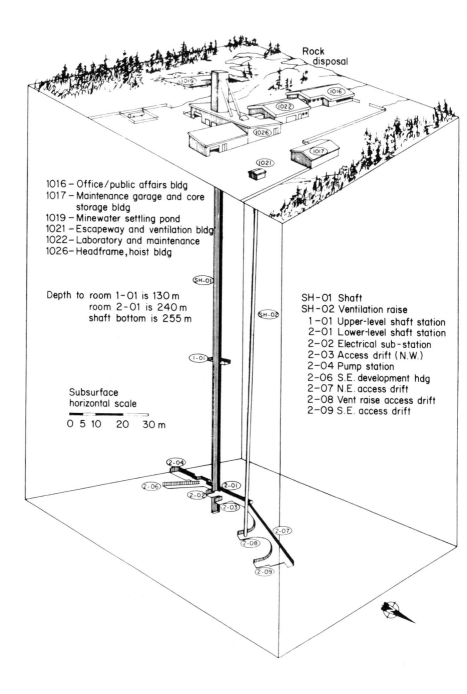

Rock disposal

1016 – Office/public affairs bldg
1017 – Maintenance garage and core
          storage bldg
1019 – Minewater settling pond
1021 – Escapeway and ventilation bldg
1022 – Laboratory and maintenance
1026 – Headframe, hoist bldg

Depth to room 1–01 is 130 m
          room 2–01 is 240 m
          shaft bottom is 255 m

SH–01 Shaft
SH–02 Ventilation raise
   1–01 Upper-level shaft station
   2–01 Lower-level shaft station
   2–02 Electrical sub-station
   2–03 Access drift (N.W.)
   2–04 Pump station
   2–06 S.E. development hdg
   2–07 N.E. access drift
   2–08 Vent raise access drift
   2–09 S.E. access drift

Subsurface
horizontal scale

0 5 10   20   30 m

Figure 7.9 Diagram of the underground research laboratory (URL) in the Lac du Bonnet granitic batholith, Manitoba, Canada, currently under construction (courtesy of Atomic Energy of Canada Ltd)

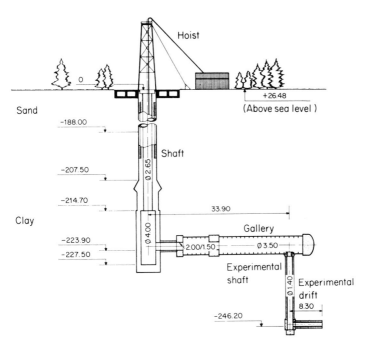

Figure 7.10 Schematic diagram of the underground research laboratory in the Boom Clay at Mol, Belgium (courtesy of CEN/SCK, Mol)

There is considerable similarity in the types of experiment in each of these URL. They fall into four categories:

(a) those designed to investigate the thermo-mechanical response of the rock to heating by the waste;
(b) experiments to examine groundwater movement at depth in the rock;
(c) radionuclide migration tests;
(d) investigation of geotechnical rock properties and techniques of mining, waste emplacement and backfilling.

URL activities are generally characterized by considerable technical sophistication, and on-site control, monitoring and data reduction by computers situated in the mines is common. The most famous URL, and the first to apply a comprehensive and integrated experimental approach, is the Stripa test mine (NEA, 1985a) in central Sweden. Originally run as a joint project by the Swedish Nuclear Fuel Supply Company (SKB) and the US Lawrence Berkeley Laboratories (LBL), this has blossomed into a multinational enterprise sponsored by the OECD/NEA. SKB maintain executive control but a large number of scientists from several countries are now actively involved in designing

158

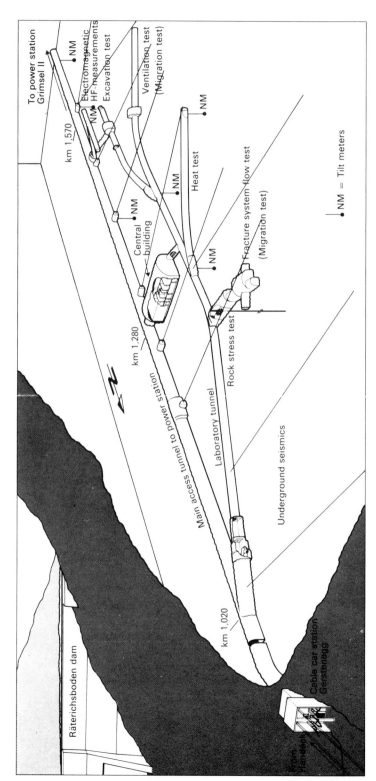

Figure 7.11 The Grimsel underground laboratory in the Swiss Alps *Reproduced by permission of Nagra, Switzerland*

and running experiments. Stripa is thus used here as an example to illustrate the four categories of experiment outlined above.

Heater experiments

Heater experiments duplicate the heat output of a canister (or canisters) of HLW using electrical heaters installed in boreholes in the rock and often backfilled as an actual disposal hole would be. Thermocouples, strain gauges and off-gas collectors are installed in the heater hole and peripheral monitoring boreholes (Fig. 7.12). Temperature can be applied in several ways: rapid rise, smooth increase, stepwise and so on. By using arrays of small heaters in several boreholes, the effect of heat on a larger zone of the model 'repository' can be assessed, and to some extent the time factor can be speeded up to simulate long post-closure times. Such experiments (Jeffrey *et al.*, 1979; Hood, 1979) have

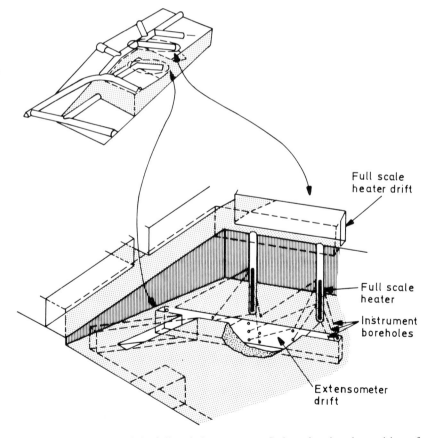

Figure 7.12 Diagram of the full-scale heater test at Stripa, showing the position of the heaters in boreholes drilled in an adit floor, and the location of surrounding instrumentation points (courtesy of SKB)

demonstrated that the simple models of heat transfer are very accurate and thermal profiles in rock can be predicted with some confidence. The effects of thermal stress on the rock have proved less predictable, particularly owing to the effects of discontinuities in fractured rocks (Chan and Cook, 1979; Stephansson *et al.*, 1979, 1980). In the crystalline rocks at Stripa, geochemical effects of heating, at the low temperatures expected in a Swedish repository, have proved to be small. Additional heater tests on buffer and backfill materials are discussed below.

Hydrogeological experiments

Various hydraulic borehole tests and deep groundwater sampling and analysis programmes have been performed at Stripa. The majority are extensions of the types of surface-to-depth borehole investigations described in the previous chapter and are not discussed further here.

In addition, experiments are underway or have been carried out to examine the permeabilities of large volumes of rock. One such experiment involved sealing off a blind gallery in the mine, circulating air through the void space thus produced, and monitoring changes in its humidity. Combined with some knowledge of the fracture pattern in the surrounding rock and of local hydraulic gradients this made it possible to calculate bulk permeability from the infiltration rate of groundwater. A current experiment monitors inflows over small areas of the walls and roof of a large gallery to assess variations in water fluxes caused by preferential pathways in a large rock volume. (Birgersson, *et al.*, 1985; Fig. 7.13).

Other experiments have attempted to link heater tests with hydraulic tests, and one trial has demonstrated the decrease in fractured rock permeability on heating, caused by expansion of the rock and consequent closure of fractures (e.g. Morrow *et al.*, 1981).

Migration experiments

In situ radionuclide migration experiments are being performed in well characterized zones of fractured rock. A single fracture transecting two galleries or boreholes can be used for such purposes (Fig. 7.14). Tracers (sorbed, non-sorbed, or colloidal/particulate species) are injected into the fracture under an imposed pressure field and their movement monitored. The results are used to validate models of lateral dispersion, sorption and pore diffusion outlined earlier. The work at Stripa has indicated that at least some tracer appears to escape from fissures into the rock matrix (Birgersson and Neretnieks, 1984) and has also shown strong indications of 'channelling' of flow within the fissures themselves (Abelin *et al.*, 1985).

Buffer/backfill behaviour experiments

An important series of experiments currently in hand at Stripa is connected with the emplacement and subsequent behaviour of a bentonite and bentonite/sand

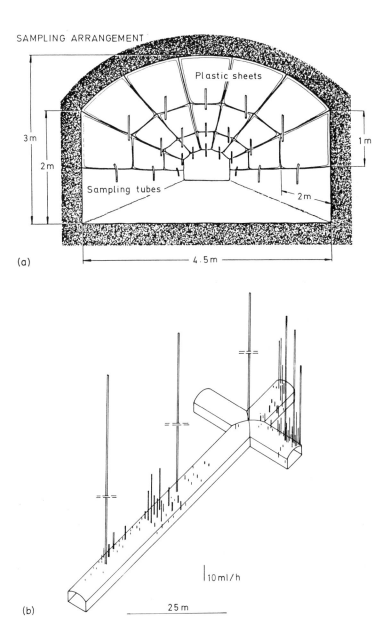

SAMPLING ARRANGEMENT

Plastic sheets

3 m

1 m

2 m

Sampling tubes

2 m

(a)

4.5 m

10 ml / h

(b)

25 m

Figure 7.13 The 3-D migration experiment at Stripa. (a) The plastic lined groundwater collection zones in the galleries. (b) Graphical results of one year of monitoring water inflow zones. The preferential pathways for water in some parts of the fracture system are clear (courtesy of SKB)

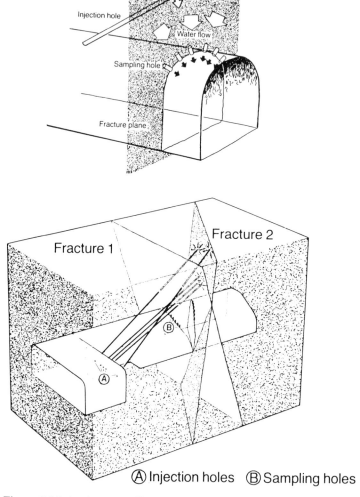

Figure 7.14 An *in situ* radionuclide migration experiment in a single fracture zone in the Stripa mine, showing the pattern of injection and monitoring boreholes (courtesy of SKB)

backfill in both disposal holes and galleries. Highly-compacted bentonite is essentially devoid of free water when emplaced but will expand on uptake of groundwater (see Chapter 5). The 'buffer mass experiment' at Stripa (Fig. 7.15) has examined a variety of factors involved in the behaviour of bentonite when rewatering and subjected to the heat output of the waste. Integral migration experiments are included and the tests are utilizing the old ventilation gallery described above under hydrogeological tests. This demonstrates the way in

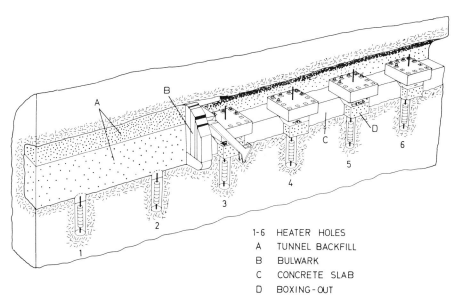

1-6   HEATER HOLES
A   TUNNEL BACKFILL
B   BULWARK
C   CONCRETE SLAB
D   BOXING-OUT

Figure 7.15 The buffer-mass test facility at Stripa (see text for details). Courtesy of SKB

which results of one experiment can be used as the starting point for another. The experimental configuration shown in Fig. 7.15 is carefully exhumed at the end of the long test period (several years) to measure the extent of rewatering, pressure injection into fissures and radionuclide movement. The results will be linked with temperature, strain and moisture content data which are being monitored remotely at various points throughout the test volume.

The experiments performed in other rock types and at other locations differ in detail, but involve similar procedures and have produced results which are broadly comparable in terms of technical level and the degree to which they have validated concepts. A good review of the present state of the art in URL experimentation is provided by Côme *et al.* (1985).

CHAPTER 8

Building and Operating a Deep Repository

The complete cycle of selecting a nuclear waste disposal site, constructing a deep repository, filling it with waste and then closing and abandoning it has never, to our knowledge, been carried through. While several deep repositories may be in operation by the end of this century, it is unlikely that any will have been completed and decommissioned before the early decades of the next century. In the case of HLW, most countries do not even envisage having an operating facility available until that time. Those countries with the most advanced programmes (e.g. Sweden, Switzerland, USA, Federal Republic of Germany) are generally at the initial stages of site selection and investigation, or are carrying out tentative pre-construction engineering and technical tests *in situ* at favoured sites. A particularly advanced example is the Swedish SFR facility (Figure 8.1), a deep repository for LLW/ILW which is being constructed at a virgin site and will be operational by 1988.

We pointed out earlier that a deep repository is essentially just another civil engineering structure and that the technical expertise to build one is widely available, requiring development only in a few specific areas such as deep, soft-ground tunnelling. In addition, all the active handling techniques necessary for waste emplacement are already available in the nuclear industry, and, as the previous chapter described, site investigation techniques exist for most geological environments. In this chapter we can therefore integrate this experience to describe, in a general way, how a deep repository might be constructed and operated. Many detailed technical designs already exist, having been prepared largely to demonstrate the feasibility of deep repository engineering. There are, of course, considerable variations in design detail, particularly from one host-rock type to another, but fundamental concepts are much the same.

Apart from technical issues, further constraints on the repository location,

164

design and operation are set by national regulations. At the simplest level such regulations may be based on political decisions to site repositories in regions which benefit from nuclear power (as in Canada) or general regulations which apply to any such engineering operation (e.g. codes of practice for mine operation). In the USA, indeed, minimum performance levels have been specified by the Nuclear Regulatory Commission (NRC) for individual engineered barriers and geosphere characteristics but the practicality and applicability of such an approach has been widely criticized. At a more general level, overall performance objectives for a repository derived from model predictions under all reasonably expected conditions have been specified in several countries in terms of maximum individual dose rates (e.g. Sweden, Switzerland) or maximum collective doses (e.g. USA). These have considerably less impact on the details of repository design than the NRC regulations, and are considered in more detail in Chapter 10.

(a)

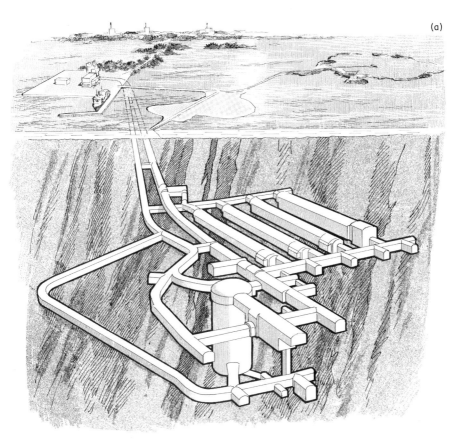

Figure 8.1 (a) The SFR facility at Oskarshamn in Sweden is a submarine repository in crystalline basement rocks for the disposal of low/intermediate level wastes. It is situated

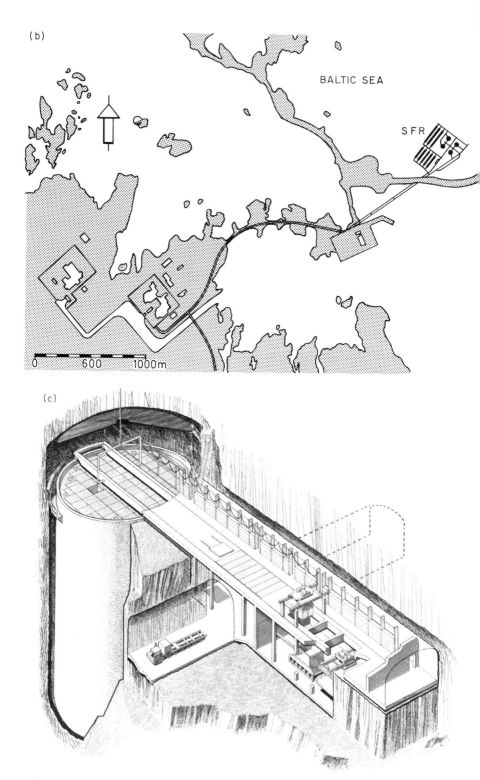

(b) just offshore from the existing nuclear power plant with access by tunnel from a small island linked by causeway to the mainland. It comprises both caverns and silos (c) for different waste types. *Reproduced by permission of SKB.*

SITE SELECTION

Selection of eventual repository sites is an issue which, ideally, should be based on a rational appraisal of all the scientific evidence available, after a long period of generic research and investigation and comparison of different geological environments and potentially suitable locations. In practice of course, as with all major developments, political and social factors define what is possible (as in the case of the proposed Billingham site in the UK), and while they may frequently frustrate the scientist or engineer, they are of considerable importance in reaching an acceptable solution.

The approach that most countries have adopted is to use the interim results of international research programmes to help define, for their own landmass, potentially suitable types of rock or, more recently and more appropriately, hydrogeologically suitable deep environments. The selection of sites for comparison within these areas is then a pragmatic choice of available land, which may be dominated by a host of factors such as availability of transport routes, local or national planning policies and land ownership.

In 1974 the IAEA began a first attempt to assemble a list of repository site selection factors (IAEA, 1977). At this stage long-term geological stability, resistance to potentially disruptive phenomena (including the action of man), and the feasibility of constructing and operating a repository in certain rocks were considered. It became apparent that many site-specific and rock-type specific factors could not be separated when assessing the potential of either a site or a host rock. This intimate relationship is highlighted by the unwillingness of any organization to define geological *criteria* for a repository which might then be regarded as widely applicable generic specifications for any site. This reflects both the highly variable nature of the geological environment and the difficulty of predicting its future behaviour.

However, the US National Academy of Science published (NAS/NRC, 1978) a brief list of geological 'criteria' while the IAEA gave a list of site selection factors which is summarized in Table 8.1 (IAEA, 1982a) and which forms as comprehensive a generic compilation as is likely to be possible. As such it is more a checklist of research requirements than stipulations for any 'suitable' site. Many countries contributed similar broad 'criteria' or selection factors relevant to their own cases (see for example, Gray *et al.*, 1976, for the earliest UK view) although all of these should be viewed as guidelines only, and were very much exercises in testing the water.

At the outset then, let us state quite clearly that there are no *criteria* in existence, nor are there ever likely to be any, which can be used universally as a basis for the selection of sites for waste repositories. Because of the unique features of any body of rock, and the complex interplay of its properties (such as fracture density, porosity, permeability and so on) all that we can do is to assess what appears on first principles to be a potentially suitable site on its own merits.

Very few features of a body of rock can be said to be universally detrimental to waste containment, each must be evaluated in the particular environment in

Table 8.1 Site selection factors for deep geological repositories (after IAEA, 1982a)

1. Topography

2. Tectonics and seismicity

3. Subsurface conditions:
   Depth of disposal zone
   formation configuration: thickness and extent, consistency, uniformity, homogeneity
      and purity of strata
   Nature and extent of overlying, underlying and flanking beds

4. Geological structure:
   Dip or inclination
   Faults and joints
   Diapirism

5. Physical and chemical properties of host rock:
   Permeability, porosity, solubility and dispersivity
   Inclusions of gases and liquids
   Mechanical and plastic behaviour of rock
   Thermal gradient: regional and local
   Thermomechanical and thermohydraulic responses
   Thermal conductivity and specific heat
   Sorption capacity
   Mineral content of water
   Radiation effects

6. Hydrology and hydrogeology:
   Surface waters: occurrence, form, volumes
   Groundwaters: occurrence, volumes, chemistry

7. Future natural events:
   Hydrological changes
   Uplifts and subsidence
   Seismicity
   Intrusions and faulting
   Climatological changes
   Topographical changes

8. General geological and engineering conditions:
   Site area and buffer zone
   Pre-existing boreholes and excavations
   Exploration boreholes, shafts, tunnels and excavations
   Spoil disposal
   Waste transport
   Engineering and construction of repository
   Operational safety and stability of repository

9. Societal considerations:
   Resource potential
   Land value and use
   Population distribution
   Jurisdiction and rights of the land
   Accessibility and services
   Other environmental impacts
   Public attitudes

which it occurs. Naturally, the process can be made a lot easier by ruling out sites which have a high probability of some known feature proving detrimental, indeed this approach was clearly adopted in the initial choice of the three rock groups for HLW disposal, discussed in Chapter 6. Rocks were ruled out generically if, for example, they commonly possessed a high permeability—which deleted sandstones, most limestones and so forth from the list at the earliest stages. However, this is not to be taken to imply that in certain environments, under specific conditions, these rocks would not prove a very suitable host for a repository for various types of radioactive wastes. The same applies for all the generic 'exclusion clauses' which have been applied; they are simply used to make the process more logical and rational. The broad-brush generic approach has proved a useful intuitive sorter, but we are now at the stage of having collected sufficient data to make more sophisticated appraisals of particular sites which allows much more flexibility in repository design optimisation. One clear example of this was the reassessment of waste containment requirements in terms of acceptable rates of release at various times in the future, based on radiological principles. This led to the slackening of the original concept of 'total containment' (i.e. forever) and assurance of very long term geological stability, which were both clearly unattainable. We now accept that the waste, or at least some of the longest lived components of it, has a significant probability of returning to the biosphere. The selection of a suitable site takes this into account, and the geological environment is investigated to assess how it would perform as a far-field barrier or dilution/retardation system, given this type of release. The information for a particular site will be fed into predictive models to assess the performance of that site and compare it to others. The general techniques of site investigation were described in the previous chapter.

REPOSITORY CONSTRUCTION

The results of detailed site investigations would, among other things, define the most suitable volume of rock for repository construction. It is unlikely that the idealized designs discussed in Chapter 4 could slot neatly into any particular site, and the repository layout would inevitably need to be tailored to the site. An example of this is the trial repository fitting for six sites in crystalline rock carried out by the Swedes (KBS, 1983), where major vertical and horizontal fracture zones controlled tunnel lengths and configurations (e.g. Figure 8.2).

If a 'virgin site' is to be developed, rather than the extension of an existing mine, for example, then the first step towards construction may be the sinking of a trial shaft or adit into the proposed repository volume. There are many advantages to be had from this approach. The first is that it allows for much more detailed examination of the properties of the rock than is possible using boreholes. The use of *in situ* geotechnical tests during excavation of a shaft will provide very precise data for use in, say, the exact design of tunnel supports in a stiff clay unit (Fig. 8.3). Second, it allows for an underground experimental

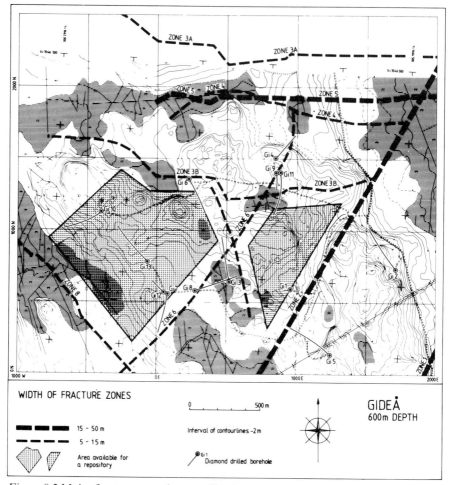

Figure 8.2 Major fracture zones in crystalline basement rocks of the Scandinavian shield would control the design and positioning of a deep repository in this conceptual study, based on real field data on fracture distributions, carried out as part of the KBS-3 project (courtesy of SKB)

phase, if this is considered necessary, in which experimental verification of any issues central to the repository performance and waste containment models can be carried out in the real volume of rock proposed for a repository. Various schools of thought maintain that some form of underground research laboratory is essential to completion of the basic research programme, but that such a chamber need not necessarily be developed into a repository, or even be in the same area. Both scientific experiments and trials of handling and emplacement techniques in underground 'demonstration' facilities, while being very expensive, can be useful means of exhibiting the technology of waste disposal. This approach is integral to the current US programme for disposal of HLW.

The third potential value of a trial excavation is that it permits further

Figure 8.3 Heavy cast iron, steel or concrete linings are required to support deep tunnels in semi-plastic clays; the trial cast iron lining of the experimental gallery at the Mol underground laboratory. The circular plate gives access to the clay, or in a repository could be the top of a waste disposal hole (see Figure 4.8). Courtesy of CEN/SCK, Mol

geological exploration of the proposed repository volume by drilling from under the ground, hence avoiding penetrating the rock with surface-to-depth holes which may be potential leakage paths back to the biosphere. Drilling horizontal boreholes from the bottom of a shaft in advance of tunnelling will give prior warning of any previously undetected geological features which may require reshaping of the repository design. The drilling of holes radial to any or all of the excavations will provide valuable additional data for the flow model which describes the performance of that specific site.

Conversion of a trial shaft into a full-scale repository is a considerable leap which would be subject to stringent licensing requirements. After completion of the trial excavations, sufficient data should have been obtained to complete a thorough safety analysis or environmental-impact assessment for that site. This analysis may form the basis for a licence to proceed with construction of the repository itself. The mechanisms for this process and the bodies involved in making the decisions will vary from country to country, and are in the majority of cases only now being defined. It is anticipated that once construction has been completed, a further application would be required before any waste could be deposited in the ground, and this would require evidence of the completed site being satisfactory from the point of view of the original safety assessment. This is parallel to the practice adopted for siting and building nuclear power stations. Depending on the size and configuration of the repository, plus any complexities in mining technique, such as the need to use full tunnel wall supports, it might be expected that the excavation phase of the operations would be completed within two to five years of a licence being granted. A simple tunnel and borehole repository is comparable in size to a small mine, and the scale of surface operations, spoil heaps, depots and associated road construction would be commensurate during the excavation phase. It should be borne in mind, however, that a repository is unlike a mine, in that mines generally follow the geometry of ore bodies and may be very complex underground warrens. A repository is a civil engineering structure designed as would be, say an underground pumped storage scheme, and would to some extent include similar complex technology associated with such ventures.

Once excavation is complete, it might be expected that the surface manifestations would be cleared up, and replaced by the much smaller scale facilities required for waste reception (and possibly packaging) and emplacement; offices, stores, road or railheads, and possibly mine lifting gear, depending on the style of underground access. The length of time these facilities are in use depends very much on the rate of production and numbers of waste packages, the size of the repository and whether there is some stipulation that the emplaced waste must be retrievable for some period. At present this latter factor is proposed for waste disposal in the USA, where a period of 50 years retrievability is required after emplacement. There is some rather muddled thought behind this requirement, since apart from the possibility of something going wrong with an uncompleted system, there is little foundation for the 50-year period, and no consideration of the circumstances under which retrieval might be considered necessary or of the relative exposures that might accrue from this operation, as opposed to any that might occur if the waste was left in the ground. In Switzerland, for example, provisions for retrievability are explicitly excluded as it is argued that disposal is final and nothing should be incorporated into a repository design which may be detrimental to long-term performance. It is to be hoped that the concept of retrieval is given much more thought as to its implications before it is built into any national legislation.

Taking the UK as an example, it is thought that one repository would be adequate to contain all the HLW arisings until at least the end of the century, and that emplacement of the waste would take place over a period of several decades. A further deep repository would be required for long-lived ILW. Since it is unlikely that a HLW repository would be available until after the turn of the century, much of the waste emplaced would be old, well-cooled material. The subsequent rate of disposal would depend very much on policy regarding pre-disposal storage periods and, of course, on the development of nuclear generating capacity over the next 40 years. The uncertainty associated with such policies is emphasised by the threats of extensive restructuring of some national programmes as a reaction to the Chernobyl accident in early 1986. It can thus be seen that no specific prediction can be made concerning the operational phase of the repository since much depends on licensing procedures, pre-disposal waste-management policy and waste disposal legislation.

OPERATION OF A HIGH-LEVEL WASTE REPOSITORY

Waste packages would arrive at the repository surface facility by road, rail or sea, depending on the siting of the repository, and would probably be in small consignments. The sequence of events which might follow is shown schematically in Figure 8.4. Off-loading of a shipment cask (the radiation shielding surrounding the sealed and completely prefabricated waste package) would take place in an active handling facility situated adjacent to the repository entrance (portal or shaft-head). The waste package would be removed from the cask remotely, and transferred either into a shielded flask on a vehicle, in the case of adit entry, or onto a lifting mechanism in the case of shaft entry. A vehicle would carry the package until it arrived at the disposal hole in the repository. A device for lowering packages down a shaft might be either shielded or unshielded, and would be designed to prevent free-fall in the case of winch or brake failure. Some designs propose the use of a narrow shaft down which a completely unshielded package could be lowered directly from the surface active handling facility, into a shielded container on a vehicle positioned at the shaft bottom in the repository. Others would use a more conventional lifting shaft with a shielded cage. In either case the waste shaft would be separate from shafts used for transfer of personnel or equipment.

Once the package is moving on the underground vehicle, it can be taken to the allocated disposal site and the package emplaced by lowering out of the shielding into the ground. Buffer material can then be placed in the annulus, and the hole or tunnel backfilled or prepared for the next package. Once a hole or tunnel is filled, a concrete plug would completely shield personnel in the adit from any radiation and operations could shift to the next one. At some stage a section of the adit might be backfilled with the chosen material (clay, rock-spoil, salt and so on) and a bulkhead wall constructed to seal off and decommission a completed section of the repository. This style of operation would continue until the repository was completely filled with waste and all the tunnels backfilled. During the years over which this takes place, consideration may have to be given to temperature rises in the rock, and to ventilation and cooling of operational parts of the repository. In many environments the repository would have to be pumped to keep it dry during operation. On decommissioning all of these processes would be halted and the remaining operating space below ground would be backfilled and bulkhead walls constructed. The final operation would be to backfill and seal all access shafts and tunnels back to the surface. This is considered in more detail in the next section.

CLOSURE AND MONITORING

When a repository is 'decommissioned' (completed and sealed) there would be no further use for the surface facilities and these too might be decommissioned and dismantled. All that would then remain after landscaping the area, is some form of permanent obelisk or plaque to mark the position of the site. Since there would be no means of access to the repository (which would no longer exist as an

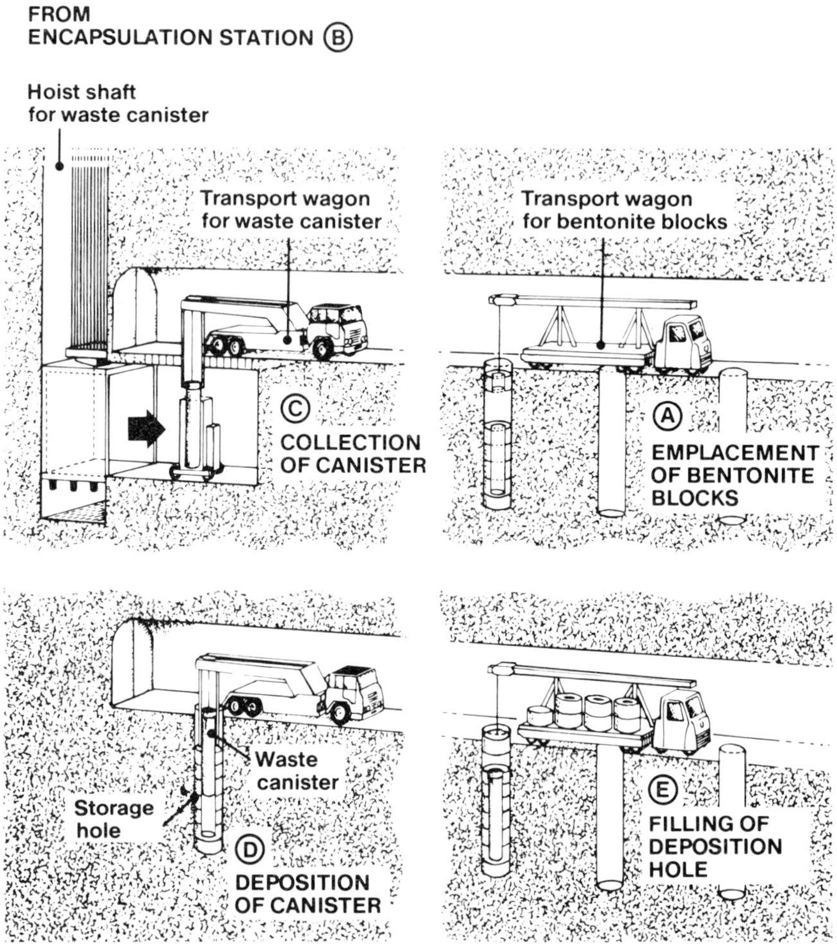

FROM
ENCAPSULATION STATION Ⓑ

Hoist shaft
for waste canister

Transport wagon
for waste canister

Transport wagon
for bentonite blocks

Ⓒ
COLLECTION
OF CANISTER

Ⓐ
EMPLACEMENT
OF BENTONITE
BLOCKS

Waste
canister

Storage
hole

Ⓓ
DEPOSITION
OF CANISTER

Ⓔ
FILLING OF
DEPOSITION
HOLE

Figure 8.4 Schematic diagram of emplacement of a high-level waste container and the surrounding bentonite buffer in a repository in hard crystalline rock (see Figure 4.3). Courtesy of SKB

opening in the ground) or the waste, there is consequently unlikely to be any need for fences or other forms of 'security' on the site.

At present there is some debate as to whether a deep disposal site should be monitored in any way after closure. The types of monitoring could take many forms, from trying to measure conditions in the rock around the waste, to monitoring radioactivity levels in local surface water bodies, if such exist in the environment chosen. The former option would involve emplacing very complex, and possibly unreliable, remote instrumentation in the repository at the time of backfilling. Communication to the surface by cables through conduits would constitute undesirable potential leakage paths. Since it would be impossible to maintain the monitoring probes, their life expectancies in the underground environment would be short, and their use consequently very limited. It is very debatable in any case whether such monitors would be able to provide any useful data on the containment of the waste, since it is clearly not possible to monitor the whole of the geological barrier for the time scales involved.

The second option superficially appears more realistic, although it may appear that one is monitoring for the effects of events of minute probability. The site performance model will have identified potential leakage routes over time scales of millenia, so surface monitoring would not be expected to detect any radionuclide movement. Bear in mind that the operators anticipate some degree of mobilisation and dispersal of the waste anyway. If releases were eventually detected many thousands of years hence what in any case could be done about it? Exhuming the waste would expose more people to immediate risk than leaving it alone.

It seems fair to say that any form of long-term monitoring of a deep-disposal system has no scientific rationale and no value in terms of any type of site safety assessment. The closure of a deep repository should be regarded as a final act, the end of operations at the site. The land surface above the repository could then be returned to its normal agricultural or other uses.

SEALING SHAFTS AND BOREHOLES

An unsealed access shaft to a repository could pose a serious potential leakage path back to the surface. Depending on the environment under investigation, one or more exploratory boreholes might also be expected to penetrate the repository rock volume, although not necessarily the workings or the disposal holes themselves. A considerable amount of research is being carried out to evaluate techniques for sealing shafts and boreholes and, in addition, the use of such methods for placing bulkhead dividers in the repository workings to separate completed sections from operational ones.

Many materials are available, ranging from local material such as crushed rock spoil mixed with cement grout or clay, through other 'natural' materials such as swelling clays, compacted soils, bitumen, pozzolans and finely crushed basalt which is hydrothermally self-cementing, to synthetic materials such as cements and grouts, resins, and pulverised fuel ash (PFA). Many concepts for seals use combinations of these materials, either in a particular seal, or in series of seals.

In a shaft it is naturally easier to provide quality assurance as the seal is accessible both during and after emplacement, and there is little doubt that series of seals would be used in this situation. The geometry of the seal is flexible (Fig. 8.5) and it may be useful to enlarge and clean the hole to be sealed by reaming to ensure both a good contact and a firm seat for the seal. This technique produces seals which are larger in diameter than the hole in which they are emplaced. Similar reaming techniques might be used to seal exploratory boreholes.

The principal problem in ensuring a good seal is not the achievement of good seal properties, for example in terms of very low permeability, but in assuring

176

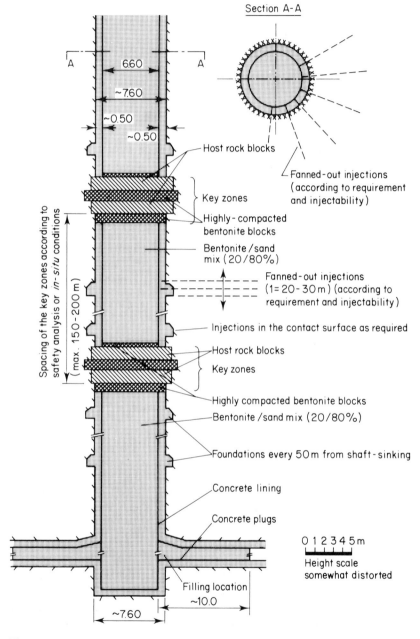

Section A-A

660

~7.60

~0.50

~0.50

Host rock blocks

Fanned-out injections
(according to requirement
and injectability)

Key zones

Highly-compacted
bentonite blocks

Bentonite/sand
mix (20/80%)

Fanned-out injections
(1=20-30m) (according to
requirement and injectability)

Injections in the contact surface as required

Host rock blocks

Key zones

Highly compacted bentonite blocks

Bentonite/sand mix (20/80%)

Foundations every 50m from shaft-sinking

Concrete lining

Concrete plugs

Spacing of the key zones according to
safety analysis or *in-situ* conditions
(max. 150-200m)

0 1 2 3 4 5 m

Height scale
somewhat distorted

Filling location
~10.0

~7.60

Figure 8.5 Depiction of the complex multiple seal techniques which might be used
to close access shafts to a deep repository (courtesy of Nagra)

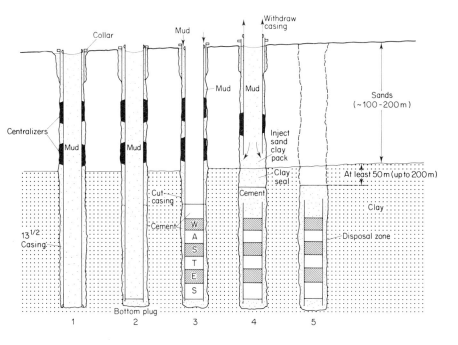

Figure 8.6 Deep borehole disposal of long-lived wastes. Schematic sequence of events in constructing a deep borehole for disposal in clay, emplacing the waste and sealing (after Chapman and Gera, 1985). This technique is currently being studied in Italy

good bonding to the rock. In field trials of seals commonly used in the oil industry the bulk of the leakage detected was by flow along the interface of rock and seal. In addition the damage caused to the wall rock by drilling or blasting can enhance permeability and allow increased flow past a seal.

The quality of a seal bond depends very much on the rock-type in which the seal is emplaced. The most problematic are undoubtedly the argillaceous rocks, in particular stiff to plastic clays such as those under study in Belgium and Italy. Bonding to clays has always been a problem in the oil industry, even using the more advanced expansive oilwell cements. The use of standard oilfield cements (usually based on Portland cement—OPC) in clays in the context of radioactive waste disposal research is quite new and recent studies (e.g. Milodowski et al., 1985) demonstrate the poor results which are normally achieved in borehole cementing. Present proposals for deep borehole disposal of HLW/ILW in such clays in Italy (Fig. 8.6) are very dependent on being able to find a seal which can be pumped into position at depths of several hundred metres, and which will set to form an interlocking mineralogical structure with the host clay. Several OPC based materials, with various additives, have been tested under appropriate *in situ* stress conditions in the laboratory, with varying degrees of success (Chapman *et al.*, 1986a). It is probable that a mixed host-clay/compacted bentonite seal will have to be developed, as this would have the most similarities with the surrounding clay.

Highly compacted bentonite has been very extensively tested for use in a variety of sealing and plugging applications in crystalline rocks. Pusch has performed several full-scale tests of adit sealing, disposal hole backfilling and borehole plugging at the Stripa underground laboratory in Sweden (Chapter 7). The purpose of these studies (e.g. Pusch *et al.*, 1985) has been to test emplacement techniques, to assess the degree of homogeneity of large volume seals, to examine the rate of uptake of water by the bentonite and the coupled effects of rewatering while under thermal load from the waste.

During the site investigations for the Swedish SFR repository, several boreholes had to be drilled into the rock beneath the sea floor, using a small jack-up platform. These submarine boreholes were subsequently sealed using the technique developed by Pusch at Stripa, whereby highly compacted bentonite was emplaced in the holes in a perforated metal tube, allowing rewatering and expansion out into the annulus and against the borehole wall.

Apart from the Stripa work, the only other large seal test which has been performed is in the Hope salt mine, near Hannover in Germany. A tapered bulkhead seal comprising various materials including 'saltcrete' (crushed rock-salt and cement) and bitumen was emplaced in a gallery in the mine and equipped with pressure and strain transducers. The mine was then flooded (as part of a larger programme to assess processes in the event of shaft seal failure in an operating salt repository), and the seal performance monitored as pressure on the flooded side increased. The trial, still underway, has proved successful so far.

The *longevity* required of a seal and its bond is, however, the fundamental issue, and it raises a more basic point regarding which criteria should be used to specify requisite seal performance. For example the actual potential of exploratory boreholes for contributing to local groundwater flow perturbation, and to increased likelihood of short-circuiting, has not been fully investigated by modelling. The effect of open boreholes, or boreholes plugged so as to achieve certain permeabilities, must be studied for specific geological and flow environments and their consequent significance quantified. It may then be possible to apply realistic numbers to factors such as requisite permeability, seal lifetime and admissible deterioration with time, which take into account the overall performance of the repository. Until this is achieved it will be very difficult to design effective seals for either shafts or surface connected excavations, without trying to achieve the impossible and unnecessary goal of a perfect seal for all time. At present it seems that the materials technology and the methods for seal emplacement are all available, although research and further development will undoubtedly be required for specific circumstances.

The general problems of sealing all types of opening were summarized at a workshop held in the USA in 1980 (NEA, 1980), and apart from the very advanced work on bentonites mentioned above, little progress has been made since that time, largely owing to lack of practical *in situ* studies. Lake *et al.* (1985) give a useful summary of the state of research in Europe which covers most of the rock formations of interest in waste disposal, and highlights the review level at which most of the work currently stands.

CHAPTER 9

Shallow Burial of Low-activity Wastes

Burial of low-level radioactive wastes in trenches or pits at shallow depths has been practised for the last 40 years, principally by those countries with early development of nuclear programmes. The volumes of low activity wastes currently being generated are many orders of magnitude greater than high-level wastes, but because of their short-lived nature and relatively low potential hazard, a simple technique such as shallow burial has generally been considered an appropriate solution to their disposal.

Burial in simple trenches up to 10 m deep with an earth or clay cap is a widespread practice. By 1980 about 760 000 m^3 of waste had been disposed of in this way in six of the principal commercially operated sites in the USA (Brookins, 1984). By 1980 the USA was generating more than 90 000 m^3 of LLW a year, most of which was being buried at the Barnwell facility in South Carolina. It is estimated that more than 10 million m^3 of LLW will have been produced in the USA by the late 1980s. In the UK the Drigg site in Cumbria currently disposes of up to $10^5 m^3$ a year in shallow trenches and additional sites are being sought for disposal of these wastes (see Appendix II).

The wastes buried in shallow trenches are now subject to stringent controls in terms of their activities and radionuclide content (see Appendix II), and in many countries this applies also to their physical and chemical form and packaging.

Although some operational sites still dispose of unpackaged and unconditioned mixed LLW in a rather haphazard fashion, there is more impetus now towards standardized packaging or compaction, allowing for better engineered disposal. Such packages are carefully emplaced in a trench, void spaces filled with earth, and the trench compacted and capped. Exact criteria for shallow land disposal are still under discussion in the USA and elsewhere and the IAEA has issued general guidelines for safe disposal operations (IAEA, 1981). It is quite obvious, however, that if such criteria and guidelines and are only now being discussed, shallow land disposal can have been the subject of only limited control in the past, particularly in the 1940's to 1960's. Materials have been buried which would now be subject to much more rigorous control, and in the early days of the nuclear industry shallow land disposal was used as a convenient technique with only limited understanding of the potential environmental effects (Stevens and Debuchananne, 1976). Consequently, there is a legacy of problems left over from those times and there have been several incidents resulting in localized releases of radioactivity, closure of sites and remedial action in the USA (Robertson, 1980).

The sites at Oak Ridge National Laboratory, Tennessee (Webster, 1979), West Valley, New York (Kelleher, 1979), Sheffield, Illinois (Cartwright et al., 1986), and Maxey Flats, Kentucky (Montgomery and Blanchard, 1979), have presented particular problems since they were developed before the importance of detailed geological characterization and groundwater surveys was appreciated. However, most of the regulations now governing trench disposal were in force by the mid 1970's and present site selection and operation procedures for shallow land disposal are the subject of more adequate methodology and safety precautions (eg. IAEA, 1982b). Clearly the history of shallow land disposal is long and complex, and it is our intention here simply to review the current practices and proposals for shallow and deep (up to 30 m) trench disposal, the techniques for selecting and developing sites, and the factors that have to be considered when carrying out safety assessments.

As for deep disposal, the principal mechanism whereby wastes buried in trenches may be returned to man's environment is groundwater transport (Papadopolous and Winograd, 1974; Duguid, 1979). As radionuclide transport paths to the surface are relatively short, hydrogeological factors are of great importance in selecting a site and designing the engineered barriers of a shallow disposal facility. In addition, because the wastes are buried close to the earth's surface, a number of other release routes become important, and these are also discussed in the next section.

DESIGN OF TRENCH STRUCTURES

The detailed design of trenches varies considerably depending on the types of waste emplaced and the geology and hydrology of the host rocks. Essentially, however, two types of trench can be distinguished: *simple* trenches which are used for LLW containing primarily short-lived radionuclides, and *engineered* trenches for short-lived ILW or LLW with higher contents of long-lived or alpha-emitting radionuclides.

Simple trenches are generally only up to about 10 m deep (Fig. 9.1), often excavated in soils or poorly consolidated sediments and constructed without any sophisticated lining or backfilling material. Simple systems of drainage may be installed in the trenches, or they may be mounded over and capped by an impermeable material such as puddled clay or bitumen, to direct rainwater away to surface drains. Waste may be emplaced in simple packages (e.g. steel drums or

182

Figure 9.1 Disposal of very low level wastes in unlined trenches; in this case slightly activated components of a large experimental rig buried at the Harwell research laboratories, UK. Reproduced by permission of UKAEA

casks) but lower-activity wastes or, for example, large items of contaminated equipment, may simply be dumped without any form of packaging.

Engineered trenches are usually somewhat deeper (some tens of metres), and are more complex. The trench is lined with a material carefully chosen for its mechanical and hydraulic properties (e.g. compacted clay, concrete), which may be specifically modified to minimize degradation due to interaction with the host rock (Fig. 9.2). Waste packaging may be quite sophisticated, involving an embedding matrix for the waste material, a container and backfill. A range of waste types may be incorporated into standard packages (e.g. concrete casks or chests) both to ease handling and to optimize use of available space. Trench backfills, caps and seals may involve a range of materials to ensure desired properties of mechanical strength, low permeability and chemical stability, and may use clay, concrete, bitumen and plastics, either individually or in specific combinations.

France, for example, has a well-developed facility for both engineered deep trenches and shallow burial at Centre Manche (Faussat, 1985), and details are shown in Fig. 9.3. It is not possible to give a comprehensive guide to which types of trench are used for which wastes. In general, simpler and cheaper designs will be used where possible, but the deciding factor will always be that the disposal option selected should satisfy radiological protection criteria when a performance analysis is carried out.

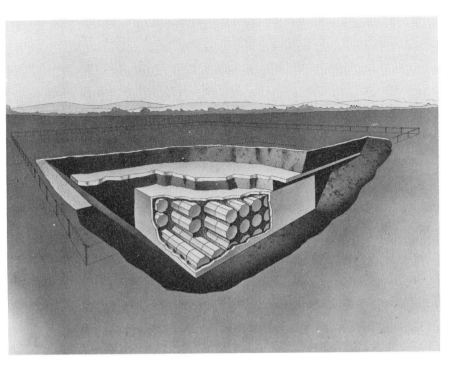

Figure 9.2 Conceptual diagram of an engineered near-surface disposal facility for some categories of low or short-lived intermediate level wastes. The packaged wastes are embedded in cement in a concrete structure, situated below a separate concrete raft designed to protect against inadvertent intrusion (courtesy of UK Nirex Ltd)

It is generally recommended practice to construct simple trenches so that their bases are above the water table, taking into account any seasonal fluctuations (Fig. 9.4). Consequently any movement of infiltrating meteoric waters is downwards past the trench, through the unsaturated zone, and into the water table. Hence any flow paths to the surface are lengthened. Depending on the physical properties of the sediments in which the trench is constructed, and in particular their permeability, it may also be acceptable to dispose of wastes below the water table. This would normally be acceptable when host-rock permeabilities are low enough to ensure that advective flow is negligible and hence diffusion is the dominant transport mechanism. Disposal in a zone through which a water table would move seasonally is to be avoided, since this leads to flooding and drying of the trenches, high potential leach rates, and high rates of degradation of the trench structure.

Most of the problems in existing sites, mentioned earlier, result from flooding of the trenches, either through fluctuation of the water table or through failure of the trench cap to deflect rainwater. Unpredictable changes in water table can be produced if extensive earth-moving during construction changes the properties of the site.

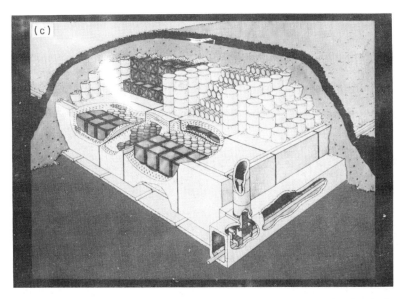

Figure 9.3 The near-surface LLW/ILW disposal facility at Centre Manche, Cap la Hague, France, showing engineered disposal of wastes in concrete containers in (a) shallow trenches capped off with more wastes (b) in a tumulus; shown schematically in (c). Courtesy of CEA/Andra

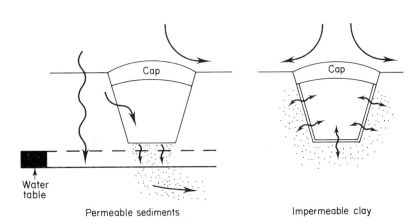

Figure 9.4 Schematic diagram of the concept of trench disposal in permeable sediments above a fluctuating water table, and in impermeable clays below the water table, where transport mechanisms are dominated by diffusion. Directions of water movement, and seasonal variations in water table are indicated (see text for details)

From the discussion above it can be seen that there are two approaches to trench siting. For disposal above the water table the host sediments should not be too impermeable as this may lead to ponding of rainwater in the trenches if an adequate cap is not provided. The ratio of cap permeability to underlying sediment permeability is thus very important, and the latter should be capable of draining percolating waters down and away from the trench while at the same time not being so permeable as to allow rapid migration of the leachate. Disposal below the water table generally means trenching in an impermeable clay. Obviously a reliable impermeable cap is essential to prevent rapid recharging and possible overflowing of the trenches in periods of high rainfall, the so-called 'bathtub' effect.

Selecting a site for simple trenches is by no means straightforward and climatic factors are clearly important, particularly when siting above the water table in humid regions. The several instances of leakages mentioned earlier clearly demonstrate the problems. However, suitable sites can certainly be found. At such sites some water is inevitably going to pass into the trenches from the unsaturated zone, so a system of gravel-pack French drains may be installed to gather this water and disperse it to underlying sediments. This prevents the wastes becoming saturated. It is possible to install piezometers and monitoring boreholes into the drain system and sediments below and around a trench to detect any migration of leached radionuclides.

Engineered trenches, being deeper, are often below the water table. Containment is ensured by a system of impermeable barriers, and any eventual release of radionuclides is controlled by diffusion in these barriers. The near-field properties are thus of considerable importance, and the models used in safety assessment must take into account gradual changes in chemistry and the hydrogeological properties of the near-field barriers as they degrade with time (e.g. see Chapter 5 for the case of cements).

Sites for trench disposal are thus selected on the basis of local and regional hydrogeology and climate. Sites overlying exploitable aquifer units at shallow depth would generally be avoided, as would areas with high hydraulic gradients or nearby small surface water bodies into which local discharge of shallow groundwaters occurs. Low-lying areas prone to flooding must be avoided and long-term ($\gtrsim 1000$ years) geomorphological changes should be taken into account. Areas of possible slope instability, high seismic risk or rapid erosion are unsuitable.

When considering the long-term behaviour of a trench site, the type of waste must also be considered. Most countries no longer bury liquid wastes of any type since rapid release may result. Consequently, some means of conditioning must be used for liquid wastes eventually bound for geological disposal. Explosive, unstable and pyrophoric materials are also excluded. Many wastes are chemically or biologically degradable and eventual gas production and settling in the trenches must be taken into account when modelling their long-term behaviour. In many older sites, where packaging of the wastes was inadequate, settling has led to problems with cracking of the trench cap and even small cave-ins. The release of biogenic gases such as methane, and gases such as hydrogen from anoxic corrosion of metals must also be considered, and for some waste types it is practice to install vents in the trench caps, or use backfills with high gas permeability. Proper packaging, emplacement, backfilling, compaction and post closure maintenance during the early life of a trench are thus essential.

RELEASE MECHANISMS FROM SHALLOW BURIAL SITES

The procedure for modelling potential doses to Man is similar to that already discussed for deep disposal of radioactive wastes. A detailed assessment of local groundwater flow must be coupled with models of the leaching and release rates of radionuclides from the near-field. The latter is very difficult to assess since, unlike HLW and many ILW types, the wastes are highly variable and inhomogeneous.

Given the great variability of the waste, complex leaching models are not generally justified for such trenches. For simple trenches, the near-field release model may simply assume a constant fractional release rate, with the entire inventory being mobilized over a relatively short period of tens or hundreds of years. Such release rates may be derived empirically from simple laboratory experiments. If the near-field chemistry can be defined (for example in engineered trenches where it may be buffered by concrete or cement), then a somewhat more sophisticated model, taking into account limiting radionuclide solubilities, can be used (as considered in Chapter 5 for deep disposal; Chapman and Flowers, 1986). If thick backfill or linings are present, diffusive transport through these can be included in the near-field model, using appropriate laboratory measured retardation data.

Evaluation of transport in the far-field is effectively identical to that for deep disposal, except that the transport paths are shorter, and possibly better defined, as are the relevant time scales. Problems in modelling near-surface transport arise primarily from complexities involved in handling the unsaturated zone, and the potentially large significance of colloids, biodegradation products and other organic compounds. For some simple trench designs, indeed, it may be possible to ignore far-field transport as such, and consider a model chain which links near-field releases directly to a biosphere model (see Chapter 10).

In addition to releases into groundwater, safety assessments must also examine other potential exposure routes which arise owing to the shallow burial depths involved. Since these wastes are of low activity, and predominantly of short half-life, they will decay to relatively innocuous levels close to the natural background within a few hundred years. However, their proximity to man's environment may necessitate some period of institutional control and restriction of access to the sites during the early, high hazard period (Pinner et al., 1984). After this period the land may be returned to normal use, with some possible restrictions. Potential exposure mechanisms during this period must be taken into account, and indeed risk analysis in a safety assessment is a valuable technique for prescribing requisite lengths of time for institutional control to minimize hazard to man.

The mechanisms which are considered can be classed as direct exhumation of the waste, for example during borehole drilling or the construction of wells or deep foundations for buildings, or surface disruption of trench seals by agricultural practice, or by civil engineering operations (e.g. road construction, pipe laying and so on). This naturally assumes a breakdown in long-term control of land use and loss of records of previous activities on the site. Some engineered trench designs now incorporate a separate 'intrusion barrier', such as a thick concrete pad located above the top of the waste emplacement vault. While this cannot prevent intrusion, it acts as a 'stop-and-think' barrier. The likelihood of occurence of any of these exposure routes can be modelled on a probability basis as described in Chapter 4. In the case of shallow trenches, agricultural use of the land and other potential biological release routes must be considered for the post-institutional control period. The mechanisms involved here are possible erosion by repeated deep-ploughing and penetration of the trench cap by deep tree roots and burrowing animals.

VARIATIONS

We have described above only the most common form of shallow land burial of radioactive wastes. There are many variations on this theme, including the use of shallow 'burial' in an engineered silo as retrievable storage for certain types of waste, which utilizes the radiation shielding properties of the ground in a cheap and convenient handling technique.

Operator exposure must always be taken into account when burying wastes, especially as the types considered here are usually only thinly shielded during transfer from transport vehicle to disposal hole. Since it may take some time to fill and cover a trench, low dose rates to those emplacing the waste containers are expected. For some types of waste the extensive surface areas involved in a large trench are unacceptable in this respect and other burial techniques are used. These include slit trenches, wide boreholes, shallow pits or concrete caissons. In each case, the waste can be emplaced so as to take immediate advantage of the radiation shielding of the earth, and the surface area of radiation 'shine' is minimized.

The use of earth-covered mounds (tumuli) to dispose of waste on the surface, while still taking advantage of underlying geology, is also a practical solution for some very short-lived wastes. In France this technique is used to cap-off deep engineered trenches. Above-ground disposal gets around the problem of saturating the wastes with water, since they are well above the water table and rainfall can be deflected by a suitable earth cap and drain system. Any leachate which would arise would permeate down into underlying sediments. Clearly such tumuli are very vulnerable to long-term geomorphological and climatic processes, so would be suitable only for the shortest-lived categories of waste.

Although it was previously observed that near-surface trench disposal of liquid wastes is rarely used now, in the past liquid wastes were occasionally poured into 'soakaway' trenches, or indeed, open ponds in unsaturated, porous

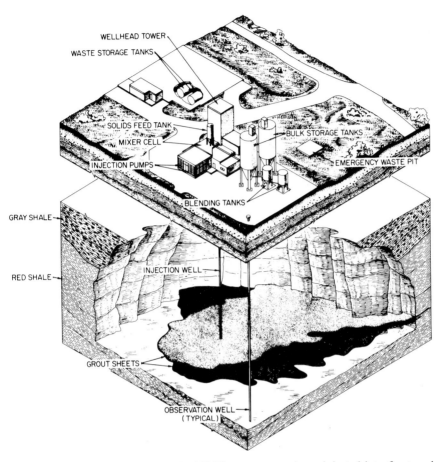

Figure 9.5 Underground disposal of ILW as a cement slurry injected into fractured shales beneath the Oak Ridge site, Tennessee. *Reproduced by permission of the International Atomic Energy Agency*

formations. A considerably more sophisticated version of liquid waste disposal is still used at a few locations. This involves mixing liquid wastes with grout to form a slurry which is injected into deep rock formations via a borehole. The rock formations naturally have to be sufficiently fractured to accept the high pressure injection, which then solidifies and immobilises the waste radionuclides in-situ (Figure 9.5). The procedure for safety analysis is similar to those considered previously, except that the near-field source term is derived simply from measured grout leaching rates (IAEA, 1983b).

A further option for shallow burial of wastes is the use of tunnels (Fig. 9.6) which might be constructed in thick sequences of clay at depths of a few tens of metres below the surface (e.g. Bechai and Heystee, 1986). Where suitable homogeneous clay deposits are available this technique can prove very attractive,

190

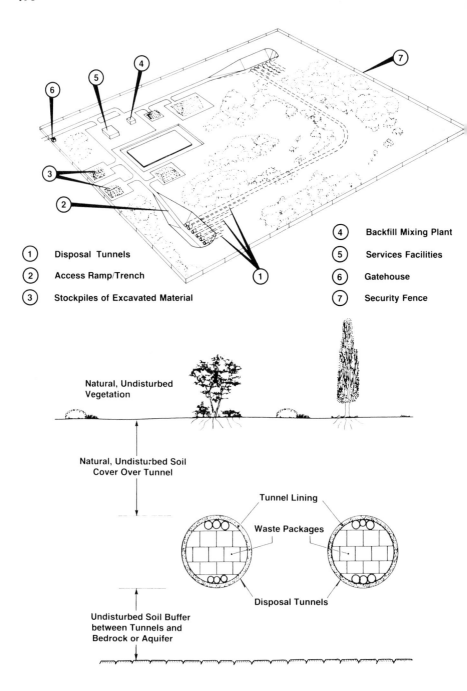

(1)	Disposal Tunnels	(4)	Backfill Mixing Plant
(2)	Access Ramp/Trench	(5)	Services Facilities
(3)	Stockpiles of Excavated Material	(6)	Gatehouse
		(7)	Security Fence

Natural, Undisturbed Vegetation

Natural, Undisturbed Soil Cover Over Tunnel

Tunnel Lining

Waste Packages

Disposal Tunnels

Undisturbed Soil Buffer between Tunnels and Bedrock or Aquifer

Figure 9.6 Conceptual design of a near-surface disposal facility for short-lived wastes using shallow tunnels (20–30 m deep) in clays (after Bechai and Heystee, 1986). *Reproduced by permission of the International Atomic Energy Agency*

both from the economic and safety viewpoint. Full-face tunneling machines can be used to excavate the tunnels relatively cheaply, and unlike trenches in clays, the material overlying the disposal zone is not disturbed by the construction process, and hence its behaviour with respect to water movement and potential radionuclide migration may be more easily predicted.

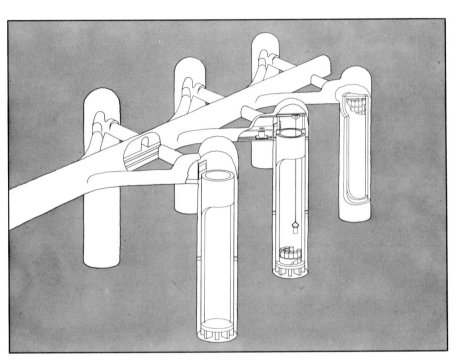

Figure 9.7 Conceptual design of concrete-lined underground silos for disposal of alpha-containing ILW in Switzerland (courtesy of Nagra)

Finally it should be noted that some countries are now opting for deep disposal of all waste types (e.g. Sweden, FRG and Switzerland). Conditioned and packaged LLW/ILW will be disposed of in large caverns or concrete-lined pits and silos excavated from underground galleries (Fig. 9.7). The rationale behind this choice was discussed in Chapter 3 and the principles of disposal and safety assessment are as described for all other types of deep burial.

CHAPTER 10

Radiological Safety Assessments

This chapter begins with a description of models of the biosphere which are used to convert radionuclide concentrations on release from the far-field into doses to man. It then reviews the purposes, objectives and scope of radiological assessments, goes on to give some examples and ends with an outline of the developments in assessment methods which are expected to occur over the next few years.

As a preliminary, we shall begin by defining exactly what is meant by the term *radiological assessment*. Within radiological protection and radioactive waste management circles 'assessment' has become a jargon word: it is used to mean both the estimation of the radiological risks associated with a particular practice (e.g. waste disposal) *and* the comparison of these risks with standards or criteria (IAEA, 1985). Thus radiological assessment is a two stage process consisting of mathematical modelling studies which provide estimates of risks, and use of these estimates to draw conclusions about the acceptability of the practices from a radiological protection point of view. In this chapter 'assessment' will be used in this sense, while 'analysis' will be used to refer to risk estimation alone.

Radiological assessments differ from the environmental impact assessments which are required by legislation in the US, in European Community countries and elsewhere. Assessments of the total impact of waste disposal on the environment needs to include considerations of aspects other than those related to radiation: for example, noise levels during operation of a repository; the presence and effects of non-radioactive pollutants in waste; the effects of repository construction on subsequent use of the site and the land surrounding it. In contrast, radiological assessments deal only with risks resulting from the radioactive nature of wastes, and are thus only one, albeit important, part of an environmental impact assessment.

CONSEQUENCES OF RELEASE: BIOSPHERE MODELLING

In previous chapters we have looked at the mobilization of radionuclides and their transport from the near-field to the biosphere. The final link of the safety assessment chain is evaluation of the consequences of any releases which are predicted, in terms of doses to the human population. This involves evaluation of processes occurring in the surface environment, for which a fairly comprehensive methodology of biosphere modelling has been developed to allow a logical approach to the problem, which is applicable to all waste types and repository systems.

Radionuclides reach the earth's surface in springs and other groundwater discharges, and are then redistributed by surface-water movements. They may finally reach man either directly, through drinking-water, or indirectly, via the food chain. Biosphere models attempt to estimate radiation doses by evaluating *dilution* in surface and groundwater bodies, and possible radionuclide *concentration* in certain biosphere 'compartments'. The first stage of the preparation of a model involves evaluation of the area in which releases are predicted to occur, in terms of groundwater and surface water classifications: springs, streams, rivers, lakes, seas and subsurface flows. For individual geographical regions a set of compartments can be defined which shows the progression of contaminants between different water bodies, along with associated dilution by uncontaminated groundwaters and meteoric waters (see, for example, Fig. 10.1). Flow between different regions, evaporation, transpiration and loss of waters with insignificant radionuclide concentrations (largely via rivers or seas) allows an overall water balance to be calculated. The area of discharge of contaminated groundwater is derived from the far-field transport model, which also yields radionuclide output as a function of time.

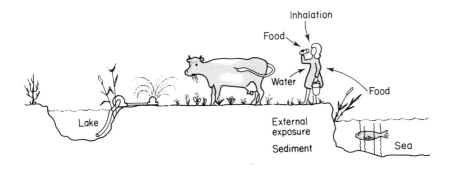

Figure 10.1 Schematic representation of the food chain and ecosystem

It is generally assumed that radionuclides are instantaneously and homogeneously distributed through each compartment of the model. This means that compartment sizes have to be chosen so that this is true over the time periods involved. Partitioning of radionuclides between water and solid phases (e.g. soils) is also taken into account, usually by use of a simple partition coefficient. It may be noted here that while underestimation of K_d values is conservative for the near- and far-field transport calculations, overestimations are generally conservative in the biosphere model, as they provide increased reconcentration within a particular compartment. The radionuclide transport rate from a compartment is proportional to its concentration in that compartment and a factor called the *transfer coefficient*.

In a second modelling stage, a food chain is constructed for each compartment. For example, a particular water compartment serves as a source of drinking-water, or crops and vegetables are planted on a particular part of a soil compartment (Fig. 10.2). When a full set of such chains has been constructed, the

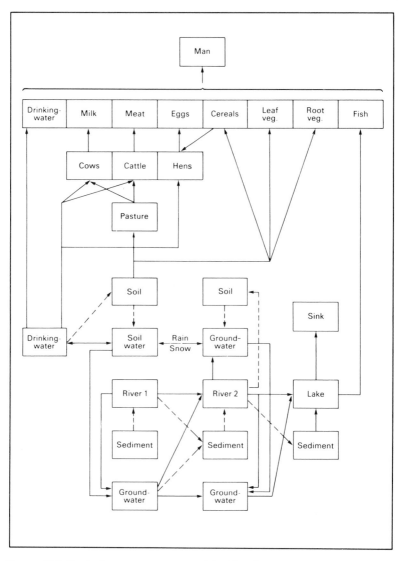

Figure 10.2 Typical compartments used in biosphere transport calculations (after Nagra, 1985). *Reproduced by permission of Nagra, Switzerland*

change in radionuclide concentration along them is calculated by the use of *concentration ratios* (CR) between various links (e.g. water to grass, grass to cow, cow to milk or beef, milk or beef to man). The calculation uses either average values for consumption of various food types, based on present lifestyles, or maximum consumption values by particular critical groups, depending on the objective of the analysis.

Although apparently straightforward, experimental determination of CR values can be quite complicated. In the last example, transfer from water to grass may be dependent on the season, and radionuclide transfer from grass to cow may vary with the type of grass, and also on whether the grass is cut and stored (as hay or silage) before being eaten. Uptake by man via various meat and dairy proucts will vary with processing and cooking. In addition, such transfer factors, like K_d values, are very dependent on the chemical speciation of the radionuclide involved. The difficulties involved in evaluation of speciation were described for the deep geological environment in Chapter 5, and can be even more pronounced in the near-surface realm, owing to:

(a) enhanced biological activity;
(b) high concentrations of a wide variety of organic compounds;
(c) marked pH and redox gradients in soils and surface waters.

For this reason, and because environmental conditions vary from site to site, it is not surprising that the reported range of variation of concentration ratios from soils to plants is large: from two to six orders of magnitude for some elements. However, it is possible to select conservative values for a particular site (e.g. Grogan, 1985).

Finally, doses are calculated from the activity concentrations in water and in the individual foodstuffs, the food quantities consumed, and *dose conversion factors* (Table 10.1). These factors are calculated using models of human metabolism, and take into account:

(a) total body mass, and the mass of each organ;
(b) radionuclide partitioning between different body organs;
(c) radionuclide retention in the body ('biological half-life');
(d) differing radiosensitivity of specific body organs.

In calculating these factors, account can be taken of variation of dose per unit intake with the age of the individual, and with the chemical form of the radionuclide (Greenhalgh et al., 1985). The doses calculated are those received over the lifetime of an individual, following one year's intake of a radionuclide, because radiological protection standards are based on the concept of limiting the risk to an individual over the rest of his or her lifetime from a year of exposure to radionuclides in the environment.

In addition to the ingestion pathway outlined above, in certain cases other exposure routes must be taken into account, such as inhalation or external

Table 10.1 Equations which are used in calculations for the BIOPATH model (after NAGRA, 1985)

General parameters

DF_{ing}	= dose conversion factor for ingestion	Sv/Bq
C_s	= concentration in soil	Bq/kg
C_w	= concentration in water	Bq/l
CMM_{xxx}	= annual individual consumption	l, kg or piece

Drinking water dose D_w

D_w	$= DF_{ing} \cdot C_w \cdot CMM_{water}$	Sv/year

Dose from milk consumption D_{mi}

D_{mi}	$= DF_{ing} \cdot CMM_{milk} \cdot DF_{milk} \cdot A$	Sv/year
DF_{milk}	= distribution factor for milk	days/l
A	$= CMC_{grass} \cdot c_s + CMC_{water} \cdot c_w$	—
CMC_{grass}	= daily grass consumption by cows	kg/day
CR_{grass}	= concentration factor soil–grass	—
CMC_{water}	= daily water consumption by cows	l/day

Dose from meat consumption D_{me}

D_{me}	$= DF_{ing} \cdot CMM_{meat} \cdot DF_{meat} \cdot A$	Sv/year
DF_{meat}	= distribution factor for meat	days/kg
A	= as under 'dose from milk consumption'	—

Dose from consumption of leaf vegetables D_{bl}

D_{bl}	$= DF_{ing} \cdot CMM_{leaf} \cdot CR_{leaf} \cdot c_s$	Sv/year
CR_{leaf}	= concentration factor soil–leaf vegetable	—

Dose from cereal consumption D_{get}

D_{get}	$= DF_{ing} \cdot CMM_{cereal} \cdot CR_{cereal} \cdot c_s$	Sv/year
CR_{cereal}	= concentration factor soil–cereals	—

Dose from consumption of root vegetables D_{wu}

D_{wu}	$= DF_{ing} \cdot CMM_{root} \cdot CR_{root} \cdot c_s$	Sv/year
CR_{root}	= concentration factor soil–foot vegetables	—

Dose from egg consumption D_{ei}

D_{ei}	$= DF_{ing} \cdot CMM_{egg} \cdot DF_{egg} \cdot (CMH_{cereal}$ $CR_{cereal} \cdot c_s + CMH_{water} \cdot c_w)$	Sv/year
DF_{egg}	= distribution factor for eggs	days/piece
CR_{cereal}	= see 'dose from cereals'	—
CMH_{cereal}	= daily cereal consumption by hens	kg/day
CMH_{water}	= daily water consumption by hens	l/day

Dose from fish consumption D_{fi}

D_{fi}	$= DF_{ing} \cdot CMM_{fish} \cdot CR_{fish} \cdot c_w$	Sv/year
CR_{fish}	= concentration factor water–fish	—

---

irradiation. Calculations involving radioactive gases, aerosols or suspended particulate material use similar compartment specifications, and intakes derived from standard inhalation rates. Resultant dose is again calculated using specific dose conversion factors. External irradiation from contaminated soil or from swimming in contaminated water can be calculated by standard techniques. Although the procedure involved in calculating doses from radionuclide release estimates seems complicated, computer codes exist which greatly simplify the procedure. For example, the code BIOPATH was used in both the Swedish and Swiss safety analyses discussed later in this chapter, and many similar codes are available (e.g. BIOS: Lawson and Smith, 1985; ECOS: Kane and Thorne, 1984; ECOSYS: ABRICOT, etc.). From a database of soil partition coefficients, concentration factors and dose conversion factors, the net doses can be calculated readily for any release scenario. Apart from the 'base case', which evaluates present conditions, alternative scenarios are often assessed. Changes in patterns of water use can be of particular significance. The drilling of wells might short-circuit some of the near-surface transport pathway and decrease the extent of dilution, while construction of dams or reservoirs could greatly increase dilution. A range of such scenarios is generally considered in a comprehensive safety assessment, and this may even go so far as to consider the effects of long-term climatic variations, for example over glacial epochs, ranging through tundra-boreal-icecap conditions. This would alter both the groundwater system and the relevant food chains. Such calculations must obviously be very speculative. If one considers the difficulty which someone in the Middle Ages would have had predicting present populations and lifestyles, it is apparent that detailed extrapolations over periods up to a thousand times longer are, at best, exercises in educated guesswork, but nonetheless necessary.

PURPOSES, OBJECTIVES AND SCOPE OF ASSESSMENTS

A full radiological assessment of any disposal method for radioactive wastes needs to consider three phases of the existence of the waste repository: the 'operational phase', when wastes are being emplaced; a possible 'post-operational, institutional management phase', when the repository has been closed but the site is still monitored and maintained; and the 'post-operational, post-institutional management phase', when the site has been abandoned.

Assessments for the first two of these phases need to consider potential radiation doses and risks, both to the workers at the site and to the general public (i.e. everyone else). For the final phase, only the public needs to be considered and the assessment needs to span much longer time scales. It is the risks during this phase, when there are no active control measures, which cause most concern and thus most of the assessments which have been performed so far have focused on the post-operational, post-institutional management period.

As indicated in Chapter 1, substantial progress has been made towards internationally agreed long-term radiological protection criteria for disposal of solid radioactive wastes. The consensus is that the objective of disposal should be to ensure that no individual, either now or in the future, should be subject to a risk from waste disposal which is greater than other risks which are widely accepted today. Furthermore, the risks to both present and future individuals and populations should be reduced to 'as low as is reasonably achievable (ALARA)', social and economic factors being taken into account. These forms of criteria, together with the worries which are often expressed about radioactive wastes, lead to the idea that radiological assessments of waste-disposal options should be as comprehensive as possible, in order to provide the maximum amount of information on which to base disposal decisions. Ideally, a radiological assessment of waste disposal should therefore have the following features:

(a) it should consider all the events and processes which could lead to release of radionuclides from the repository, or which could influence the rates of release and the subsequent risks;

(b) it should produce estimates of the risks to individuals and populations, as a function of time and space, and information about the various components of these risks (e.g. whether they are due to high probability/low consequence events or low probability/high consequence events, whether many or only a few people could be affected);

(c) it should indicate the uncertainties in the risk estimates, and identify which of the parameters and assumptions have most influence on the results;

(d) it should contain information about the radiological risks to ecosystems, as well as those to people.

As will become apparent from the examples given later in this chapter, none of the assessments of geological disposal which have been published so far have been fully comprehensive. The reasons for this are twofold: first, research into disposal of high-level wastes is not sufficiently far advanced to allow full assessments to be attempted; second, many of the assessments which have been made have been preliminary and designed to indicate research priorities or to establish whether it is worthwhile to continue to study particular disposal methods, rather than to provide a basis on which to take final decisions about waste disposal.

It must also be emphasized that there are situations in which it is not worthwhile devoting a great deal of time, effort and money assessing disposal risks (e.g. of very low level wastes). There are cases when it can be shown through simple calculations, using pessimistic assumptions, that the radiological impact of disposal will be totally negligible.

The starting point for any assessment of the long-term radiological impact of waste disposal, whether comprehensive or more limited in scope, is to list all the events and processes (natural, human-induced and waste-related) which could initiate release of radionuclides from the repository and transport them through the environment to humans, or could influence release and transport rates. These lists are generally established simply by asking people who are concerned about waste disposal to use their imagination (see Chapter 4). Once this list is available, the next step is to decide which of the events and processes are to be included in the risk analysis and how they are to be taken into account. These decisions depend on the purpose of the analysis, the resources of those carrying it out, and the amount of information available about the prospective disposal site and the behaviour of the wastes once they have been disposed of. In preliminary radiological assessments of geological disposal, the procedure generally followed is to select a few events and processes, and to combine these into a small number of 'scenarios' for risk analysis. In some of these assessments, the probability of occurrence of these scenarios is estimated, while in others the consequences of each scenario are calculated but no attempt is made to estimate the probabilities that these consequences will occur.

For a comprehensive assessment the procedure would be as follows. First, the events and processes are grouped into the following categories:

(a) those which are certain to occur (i.e. are 'normal');
(b) those which might occur and which, if they did, would perturb conditions in and around the repository but would not lead to a radionuclide release and transport scenario which is completely different from the 'normal' one;
(c) those which might occur but which, if they did, would lead to very different radionuclide release and transport scenarios, but would not lead to large, direct releases of radionuclides into the human environment;
(d) catastrophic events which could give rise to large direct releases with high consequences but which generally have very low probabilities of occurrence.

The combination of events and processes in group (a) is used to define the 'normal scenario' referred to in previous chapters. In some geological formations (e.g. salt) the consequences of this scenario will be essentially zero, that is, the most probable outcome of disposal is that there will be no risks to humans or the environment. Events and processes in group (b) can be dealt with in two ways. Either they can be included in the analysis of the uncertainties associated with prediction of the consequences of the normal scenario, or they can be combined into a series of what are sometimes called 'altered evolution' scenarios. In either case, their consequences are calculated using the same mathematical models as are used for the normal scenario but the probabilities of occurrence of these events and processes must be taken into account.

Group (c) events and processes are usually dealt with by a conventional probability-consequence approach. That is, appropriate mathematical models are developed to calculate their consequences, and then these are combined with their probabilities of occurrence to give an estimate of overall risks. Group (d) events can also be dealt with by the same means as group (c). However, group (d) events are often given separate consideration, partly because their probabilities of occurrence are so low and partly because their non-radiological consequences may well be greater than their radiological ones. (Examples of events in this class are the impact of very large meteorites and the occurrence of extensive magmatic activity, such as the formation of volcanoes.) Because such events are rarely taken into account when taking other decisions, for example whether to build cities, bridges or chemical plants, there is a growing feeling that they can legitimately be excluded from radiological assessments of geological disposal of radioactive wastes, although some of them will need to be considered when selecting prospective disposal sites.

The grouping of events and processes into categories leads to a qualitative definition of all the radionuclide release and transport scenarios which are to be included in a risk analysis. The second step in this analysis is to provide a quantitative description of each scenario, assigning numerical values to all the parameters to be used in the mathematical models and to the probabilities of occurrence of the various phenomena. Even when the available database is extensive, this parameter assignment exercise is not an automatic, completely objective procedure. In every case it is necessary to make judgements both about the 'best estimate' values of particular parameters and about their ranges or statistical distributions. For this reason it is important to identify the crucial parameters and assumptions in the analysis, so that the effects of differing judgements can be investigated.

EXAMPLES OF THE RESULTS OF ASSESSMENTS

At present there have been no fully site-specific assessments carried out which endeavour to treat all mechanisms of mobilization, transport, release and uptake realistically. All available assessments either include oversimplifications (intended to be on the conservative side) or have large generic components (such as assuming that all wastes, of all types, from a nuclear power programme are contained in one repository). In this section we examine in detail two examples of the most advanced assessments to date. These are mainly concerned with deep disposal of long-lived wastes, principally HLW. Before doing this, however, it is useful to look back at some of the earlier, and somewhat simpler, work which laid the foundations for these projects.

The generic studies of Burkholder *et al.* (1976) and Hill and Grimwood (1978) are seminal works which have contributed greatly to developing procedures for modelling repository behaviour. They contain a selection of data which is very conservative and, to some extent, arbitrary. Such generic analyses have been used primarily in the development of methodology and in defining criteria for the first stage of site selection and repository design. The Hill and Grimwood study was based on a generic inland crystalline rock site for disposal of HLW, and was followed by a sensitivity analysis (Hill, 1979), and a comparative evaluation of an equivalent repository located at a coastal site (Hill and Lawson, 1980). Results of these studies could thus be used to give some indication of the relative significance of various components of the model chain (for example, indicating the relative importance of the various near-field barriers), and to illustrate the dependence of performance on site location.

The numerical values derived from such studies must be treated with great caution as they are not intended to predict specific doses. The definition of important radionuclides in releases depends on the assumed waste inventories, which included unrealistically high ^{14}C in the case of Burkholder, and ^{129}I in the HLW case of Hill and Grimwood. These very simple models also took no account of several factors which were subsequently found to be highly significant, such as solubility restrictions and realistic redox conditions for the definition of appropriate sorption data. The models are, in fact, extremely sensitive to the choice of parameters and their values. It was shown (EPRI, 1979) that with somewhat more realistic data for the flow path, flow rate and dilution, the doses calculated for I and Tc by Hill and Grimwood are decreased by a factor of 500 and, for a period of more than 10^7 years, no significant dose would result from any radionuclide.

Despite the caveats on the numerical values produced, Figure 10.3 illustrates in a simple manner the main results of these two studies, presented as variation in resultant dose with time (from EPRI, 1979). For comparison, the results of the Swedish study of vitrified HLW disposal in crystalline rock, called KBS-1 (KBS, 1978), and a generic study by Berman et al. (1978) of a repository situated in argillaceous rock or salt, are also plotted. In addition the results of the KBS-3 and NAGRA studies, which we shall look at in more detail shortly, are included. Apart from the preliminary observation that all these studies predict releases significantly below natural background levels, it may be noted that the analyses tend to fall into two groups: those with maximum doses much greater than 10^{-6} mSv/year, and those with doses much less than this value. This can be interpreted, to a large extent, in terms of the degree of conservatism of the approach; very conservative, using highly pessimistic values for all uncertain parameters, or using realistic, though still conservative, values for parameters. The differences in these approaches are illustrated in the case studies below.

Example 1: Sweden; the KBS-3 assessment

In 1977 the Swedish government passed a law stipulating that no further nuclear reactors could be built or fuelled until it had been shown that high-level waste could be managed and disposed of safely. Since then, three major assessments of disposal have been performed by the KBS organization, which is financed by the Swedish utilities (KBS, 1977; 1978; 1983). One of these deals with vitrified high-level waste, and two with direct disposal of spent nuclear fuel. The first and third of these assessments have been subject to extensive international review, and have been used by the Swedish authorities to decide that a limited number of new reactors can be brought into operation. Since KBS-3 is the most recent of these studies, and the one which has the benefit of the most field and laboratory research, we will use this as an example.

Despite recent advances in assessment methods, the approach used in KBS-3 is very similar to that utilized in earlier studies, such as those of the NRPB referred to above. Although most of the mechanisms by which radionuclides might be released from a repository and transported through the environment are

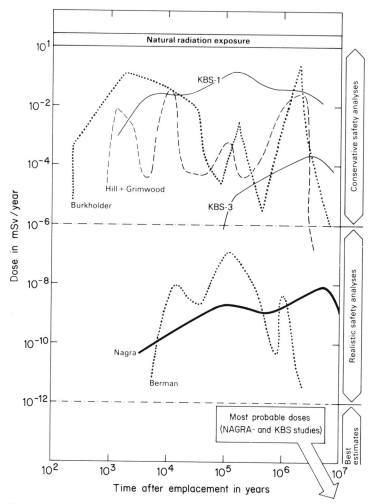

Figure 10.3 Radiation doses calculated in some important safety analyses carried out in recent years; see text for explanation (after Nagra, 1985). *Reproduced by permission of Nagra, Switzerland*

discussed in the report, calculations of potential doses to individuals and populations are carried out only for a small number of scenarios involving radionuclide movement with groundwater. No attempt is made to estimate the probabilities of occurrence of any of these scenarios or to utilize possible ranges or distributions of the parameters used in the models in order to quantify uncertainties in predicted doses. Also, in common with most other assessments, it is assumed that conditions in the biosphere remain unchanged throughout the time period considered. This assumption would be reasonable only if this period were very short, but in KBS-3 it is some 10^{10} years (twice the age of the earth!).

Having said this, however, KBS-3 does provide what is widely accepted to be a convincing demonstration that spent nuclear fuel can be disposed of in a way which presents very low risks. The models and parameters used tend to overestimate rather than underestimate doses, and the level of detail included in the models is roughly commensurate with our present confidence and ability to quantify and understand processes, and the links between them.

The most detailed part of the analysis concerns the near-field mobilization of radionuclides from the spent fuel in copper containers, surrounded by a bentonite buffer. A sophisticated model, addressing many of the processes described in Chapter 5, is built up to describe the movement of corrodants in and out of the buffer, and the rate of dissolution of container and fuel is derived. The development of a redox front in the near-field is assessed, and the transfer of key radionuclides from the near-field, across the redox front and into the far-field is calculated. At this stage the analysis makes the very pessimistic assumption that the far-field comprises only a 100 m long single fracture, which connects the repository to a major vertical fracture system where water movement is so rapid that it can to all intents be assumed to be part of the biosphere. Although realistic values of groundwater flux along the connecting fracture are used, and matrix diffusion is attributed a major role, this is nevertheless a very conservative approach to the far-field, given the amount of available data which could have been used. The biosphere part of the analysis uses the well known BIOPATH model discussed earlier and assumes groundwater releases either to a well (intersecting the major fracture zone) or to a lake of about 3 million cubic metres volume. Although the lifetime of the copper container (i.e. the initial total

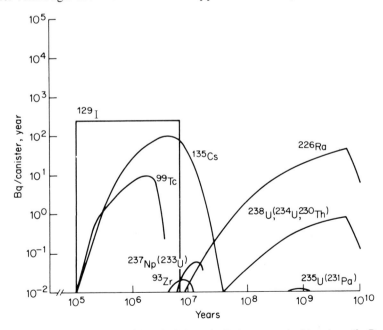

Figure 10.4 Results of the KBS-3 study. Releases to the biosphere (in Bq per container of waste per year) as a function of time after fuel discharge for the central scenario (see text). Courtesy of SKB

containment period) is calculated to be in the order of 10^7 years, a pessimistic assumption of failure after 10^5 years is made. This shows up in the predicted releases of activity to the biosphere for the 'central scenario' described above (Figure 10.4), where no release occurs before 10^5 years. Thereafter, up to 10^7 years, ^{129}I dominates releases and consequent doses. In some respects ^{129}I dose levels are an artefact of the lack of a realistic far-field component in the migration model. The 'instant' releases observed highlight the lack of any near-field retardation mechanism for iodine, although the spreading of release due to the 'mixing tank' effect of backfill and spread of canister failure times is taken into account. Beyond 10^7 years, ^{226}Ra dominates doses, but the time scale to which the analysis extends is clearly invalid, as geological and climatic conditions will not remain stable for such periods. The maximum individual dose levels for this central scenario (Figure 10.5) are 10^{-6} to 10^{-7} Sv/a, up to four orders of magnitude lower than that from natural background radiation in Sweden.

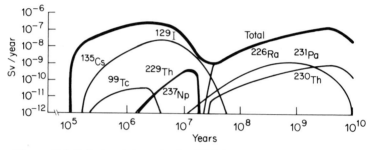

Figure 10.5 Calculated doses in the KBS-3 central scenario (see text)

Four other scenarios were analysed, including the assumption of initial damage of a single container, the prevalence of near-field oxidizing conditions which increases the mobility of certain radionuclides, transport by colloids, and concentration of released radionuclides in a peat bog rather than release to a water body with turn-over and dilution.

The effect of one of these is shown in Figure 10.6 which shows releases to the biosphere under near-field oxidizing conditions. Comparison with Figure 10.4 shows how ^{99}Tc now dominates releases, and how releases of all radionuclides are accelerated owing to more rapid breakdown of the fuel. Nevertheless, the maximum individual doses remain well below background, at only 9×10^{-6} Sv/a.

The overall results of the various scenario models are shown in Table 10.2, and doses relative to background and other factors are summarized in Figure 10.7. The implications of this very interesting assessment are discussed further in the closing section of this chapter.

Example 2: Switzerland; NAGRA Gewähr study

As with the Swedish case, the drive behind the Swiss programme was derived from government regulations in 1978 which required that safe management and

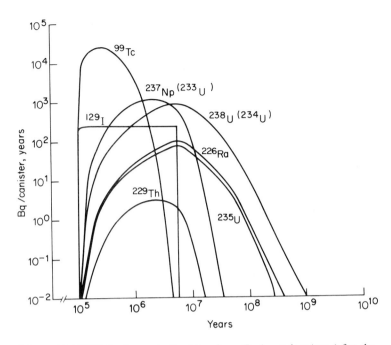

Figure 10.6 KBS-3 releases to the biosphere (Bq/container/year) for the case of oxidising groundwater conditions (cf., Figure 10.4)

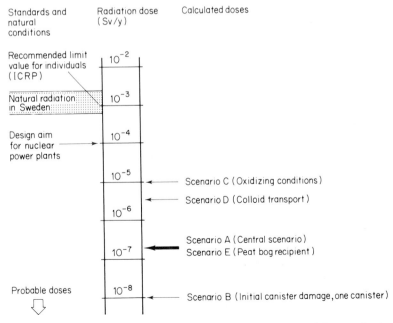

Figure 10.7 The calculated radiation doses from the various KBS-3 scenarios in perspective with natural background radiation (courtesy of SKB)

Table 10.2 Summary of the results of the KBS-3 assessment for the five release scenarios considered (KBS, 1983)

Scenario	Main characteristics	Summary of calculation results
A (central scenario)	The canisters disintegrate successively during the period 10^5 to 10^6 years. The water flow is 0.1 $l/m^2 \cdot$ year). Reducing conditions in the rock. Migration distance 100m	The matrix is dissolved in 7 million years. Uranium, neptunium and technetium precipitate at the redox front. The release of uranium to the geosphere takes 10^{10} years. The releases of uranium and radium to the biosphere are retarded until after 10^8 years. Neptunium is retarded so that it decays for the most part in the rock. ^{129}I and ^{135}Cs dominate the doses before 10^7 years, after which ^{226}Ra is dominant. The maximum dose level is around 10^{-7} Sv/y
B (initial canister damage)	Largely the same conditions as in A, but one canister is assumed to leak after 10^2 years	The releases per canister of the uranium isotopes radium, neptunium, caesium and technetium are nearly identical with A. Plutonium and americium are retarded so greatly that they decay to negligible levels before they reach the biosphere. The pulse releases, during about 2 000 years, of 10% easily accessible ^{129}I and ^{135}Cs dominate the dose. The maximum dose level from one canister is 10^{-8} Sv/y
C (oxidizing conditions)	Same as in A but oxidizing conditions in the rock, which affect solubility and sorption for uranium, neptunium, plutonium and technetium	The release to the geosphere increases for uranium (factor of 10^3), neptunium (factor of 10) and technetium (factor of 10^2) compared to A. The same elements are also sorbed less during their migration in the rock, which leads to 10^4 times higher releases of ^{237}Np than in scenario A. Maximum dose level is 9×10^{-6} Sv/y
D (colloid transport)	Irreversible sorption on particles (concentration 0.5 mg/l) in proportion to the concentration of the radionuclides in water and their sorption coefficients (K_d). Transport of the particles without retardation in the rock.	For the actinides, the particle-bound activity constitutes about 0.25% of the quantity of actinides in the water. The release of ^{231}Pa increases considerably and ^{231}Pa becomes the dose-dominating nuclide. The maximum dose level is 4×10^{-6} Sv/y

Table 10.2   Summary of the results of the KBS-3 assessment for the five release scenarios considered (KBS, 1983)—*Continued*

Scenario	Main characteristics	Summary of calculation results
E (peat bog recipient)	Same input data as in A but the biosphere recipient is assumed to be a peat bog that is used as soil conditioner after 10 000 years	The dose from ^{226}Ra is somewhat higher than in A. The total dose is the same as in A

disposal of radioactive waste should be demonstrated before operating licences for further nuclear reactors were granted. An extensive safety assessment called Project Gewähr (Guarantee) was published at the beginning of 1985 by the Swiss national cooperative for the disposal of radioactive waste (NAGRA). This analysis is extensively based on the previous Swedish KBS studies but is considerably extended in content, especially in its additional treatment of LLW/ILW types. Two repository types were considered in detail—a HLW repository in granite bedrock overlain by thick sediments, and a deep LLW/ILW repository in a marl.

The HLW safety assessment examines in detail only the disposal of vitrified waste, but the option of the direct disposal of spent fuel is not precluded. Some of the highest actinide content ILW would also be included in the HLW repository. However, due to its small effect on total repository activity inventories, this category was not explicitly included in the safety assessment.

The NAGRA HLW concept was chosen as an example of the near-field in Chapter 5 and was discussed there in some detail. Following on from that discussion, we can show how the numerical results in this safety analysis are obtained. A simple near-field release profile for Swiss HLW was presented in Chapter 5 (Figure 5.8) and we can see how this is transformed by migration through the far-field in Figure 10.8. It is evident that the processes of dilution, dispersion and decay during transit have greatly reduced radionuclide concentrations and spread their releases over longer times. Analysis of the consequences of such releases into the biosphere leads to dose predictions for each of these radionuclides arising from a number of different food chains (Table 10.3). This model also gives the total dose resulting from each radionuclide component as a function of time (Figure 10.9).

It should be noted that final doses from the NAGRA study are extremely low, and are in fact much less than those from the KBS studies. Although there are some technical factors involved (relating to the greater depth of the NAGRA repository, potential site geology and so on), the main reason for the large differences is that the KBS studies are designed to be very conservative (particularly in their treatment of the far-field), while in the NAGRA study the base case is as realistic as possible, though still tending to conservatism. The

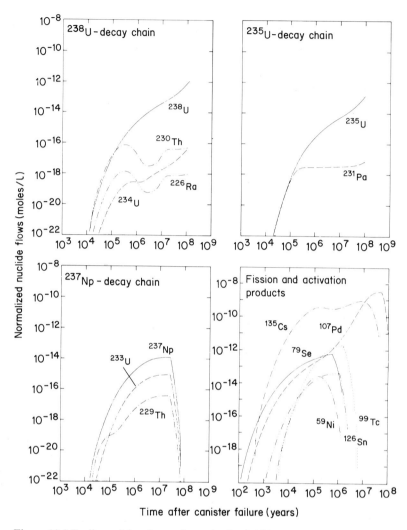

Figure 10.8 Radionuclide releases from the far-field into overlying aquifers as a function of time after container failure (from Nagra, 1985). *Reproduced by permission of Nagra, Switzerland*

Swiss study also involves a fairly extensive examination of the effects of parameter variations, and small changes in the conceptual models of the near-field, far-field and biosphere.

The LLW/ILW repository study investigates deep disposal of all wastes arising from the current Swiss power programme, plus those from medicine, industry and research produced over a period of 70 years. Although sites in crystalline rock and anhydrite are also under investigation, the study is based upon a repository in marl beneath a hillside with an overburden of $\sim$ 450 m.

Table 10.3   The distribution of radionuclide ingestion doses over various pathways for the Swiss Project Gewähr (NAGRA, 1985). The values are given as a percentage of the total dose from each radionuclide, and apply to the biosphere transport reference case used in the assessment

Exposure path	Drinking water	Milk	Meat	Leaf vegetables	Cereals	Root vegetables	Eggs	Fish
Nuclide								
^{237}Np	84	(—)	(—)	1	2	13	(—)	(—)
^{233}U	96	1	(—)	(—)	2	1	(—)	(—)
^{229}Th*	30	(—)	1	6	28	35	(—)	(—)
^{238}U	96	1	(—)	(—)	2	1	(—)	(—)
^{234}U	96	1	(—)	(—)	2	1	(—)	(—)
^{230}Th	27	(—)	1	6	29	37	(—)	(—)
^{226}Ra*	64	1	1	1	25	8	(—)	(—)
^{235}U	96	1	(—)	(—)	2	1	(—)	(—)
^{231}Pa	25	(—)	27	4	7	37	(—)	(—)
^{227}Ac*	78	(—)	21	(—)	(—)	1	(—)	(—)
^{59}Ni	86	3	2	1	5	3	(—)	(—)
^{79}Se	30	3	66	(—)	(—)	(—)	1	(—)
^{99}Tc	21	54	1	1	15	8	(—)	(—)
^{107}Pd	86	3	2	1	5	3	(—)	(—)
^{126}Sn*	12	5	(—)	3	63	17	(—)	(—)
^{135}Cs	37	22	23	3	8	7	(—)	(—)

* Short-lived daughters taken into account.
(—) Contribution less than 0.5%.

Fewer data are available on the reference site than for the HLW case (as no field site investigation for the former had then been initiated) and hence the approach tends to be more generic and more conservatively based. Although a variety of waste matrices and canisters are used, the analysis takes into account only diffusive resistance in the concrete/cement overpack-filling-tunnel liner, without considering, in the base-case, elemental solubility limits in this region. Far-field transport is conservatively assumed to occur predominantly in fissures without matrix diffusion playing a significant role. Although base-case analysis indicated low release rates, extensive parameter variations showed that significant doses could result under some circumstances. Values comparable with (yet still usually below) the safety limits set were calculated for certain human intrusion or pessimistic erosion scenarios. The fact that higher doses were predicted from the LLW/ILW repository than from the HLW repository can again be attributed mainly to the extreme conservatism of the former.

As with KBS-3, the NAGRA study does not consider some factors which could perturb the mechanistic basis of the calculation chain, such as the role of colloids, micro-organisms and organic radionuclide complexing agents. Such limitations are acknowledged in the Gewähr report and are the focus of current research programmes.

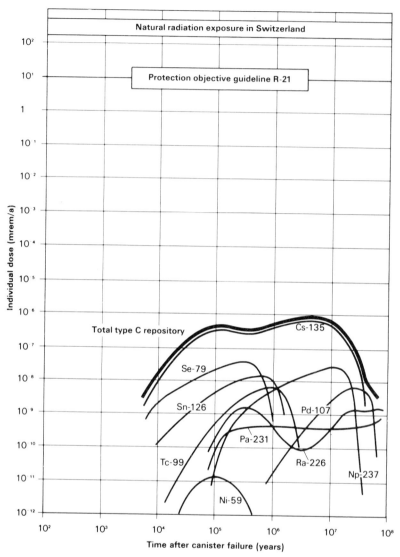

Figure 10.9 Annual individual doses from releases from the HLW repository (Type C) as a function of time for the Nagra biosphere base-case (see text) (after Nagra, 1985). *Reproduced by permission of Nagra, Switzerland*

PROBABILISTIC ASSESSMENTS

All of the assessments described so far are essentially *deterministic*. That is, they use the normal case model discussed throughout this book, and use the most likely parameter values, plus educated guesses as to the upper and lower limits feasible for those parameters. These are often brought together in some kind of simple sensitivity analysis.

More sophisticated ways of dealing not only with processes whose parameter values may vary quite widely, but also with events which have certain likelihoods of occurrence, is to carry out a *probabilistic* assessment. In this case, single parameter values are not used, but instead a probability distribution is drawn up for that parameter, which includes a range of the potential values. The calculation then randomly selects from such distributions for each parameter required, and may even randomly select events which may perturb the normal case model.

A large number of calculations can be carried out and the results can be presented very simply, for example as the probability of any given dose occurring, or the probability of it occurring at any given time in the future. Such results are shown in Figs. 10.10 and 10.11, which are drawn from the Canadian research programme on spent-fuel disposal in crystalline rocks. Atomic Energy of Canada Ltd. (AECL) have developed a probabilistic modelling system based on their Systems Variability Analysis Code (SYVAC), which is the best known of

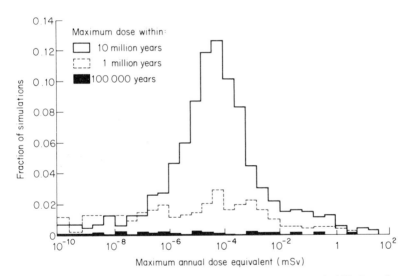

Figure 10.10 Typical results of the Canadian SYVAC probabilistic safety assessment code, run for a HLW repository in crystalline basement rocks (from Goodwin, 1985). Maximum doses which result from each simulation are plotted in the form of a histogram. A single simulation takes randomly selected values from a range for each parameter involved in the complete release and migration model used in the code. The histogram thus shows the most probable performance of the disposal system. It can be seen that for times up to 10^5 years none of the simulations leads to doses greater than natural background (around 1 mSv), and at very long times into the future (10^7 years) the most probable doses which might result are around 10^{-4} mSv, about the same as those calculated in the KBS-3 study; see Figure 10.7. Even at such very long periods into the future the probability of doses exceeding background are very small, but in devising a disposal system it is clearly worth considering the factors in the specific simulations which give rise to these doses very carefully

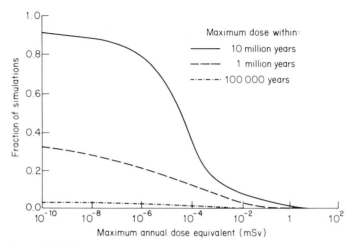

Figure 10.11 The cumulative fraction of simulations (see Figure 10.10 for explanation) which give rise to various maximum doses at three times in the future. For example, the probability of receiving a dose of one millionth of natural background (i.e. 10^{-6} mSv) at 10^5 years after disposal appears quite small; less than 5% of the simulations result in such doses. One million years after disposal some 20% of simulations reach this dose level, and after 10 million years 80% of simulations produce such doses. It can be seen that the chances of receiving doses approaching background level are small at any time. For example, even after 10 million years only about 10% of simulations give rise to doses of 1% of background

its type in the field (Dormuth and Sherman, 1981). Considerable interest is currently being shown in such codes and the CEC now has its own probabilistic code under development at the Joint Research Centre at Ispra, in Italy. This is called LISA (Long-term Isolation Safety Assessment; Saltelli *et al.*, 1984).

While probabilistic models have the advantage of looking at many possible scenarios, the large number of calculations that must be performed means that individual calculations need to be as simple as possible. This is achieved by forming a complete system model from a group of subset models (e.g. of leaching mechanism, diffusion through buffer, speciation, etc.). Thus one disadvantage of probabilistic models is that we must sacrifice some of the sophistication which is possible for many of the processes considered, while the second is simply the lack of data on parameter value distribution functions, which is particularly critical at present. Clearly, future development must aim at producing the most simple but realistic submodels, backed-up by comprehensive data distribution sets.

THE FUTURE

In recent years much emphasis has been placed on the linking, or coupling of models. This is an issue which is clearly of fundamental importance to both deterministic and probabilistic assessments. As will have become clear in Chapters 5 and 6, each process which occurs during the mobilization and transport of radionuclides can affect several others, which in turn produces additional effects which can loop back to affect the original process. Thermal effects control geochemical interactions which can alter the mechanical stability of the host-rock, which can affect their heat transfer behaviour, and so on. There are many examples of this.

Work is now progressing towards 'fully coupled' models, which take many processes and feedback loops into account. This is a very tall order, not least from the viewpoint of the computational power required. For instance, at present even the biggest supercomputers available are barely capable of handling, with realistic run times, the simple coupling of geochemical equilibration codes with those for two- and three-dimensional groundwater flow. Consequently it is important to evaluate whether coupling is essential or not, and we must always seek simplifying assumptions, while at the same time being able to justify their use.

A current development is seen in the move away from generic studies towards more site-specific work. While it may still be argued that some components of the near-field can be modelled without detailed knowledge of the repository site, it is becoming increasingly evident that far-field transport processes can vary considerably in their relative importances from site to site, even within a given geological environment. As potential repository sites are identified and the process of characterization begins, detailed information becomes available to allow more sophisticated models to be developed which simulate radionuclide release and transport more realistically. Although such models will contain simplifications and extrapolations from laboratory data, they can be validated by *in situ* tests at the site to which they are being applied, and possibly by *in situ* natural analogue studies.

As models are developed, and the associated computer codes become more complex, they must be verified by a continual process of intercomparison. Apart from comparison of individual submodels (see Chapter 11), comparison of the complete model chains used for assessment is also required. At present the CEC is carrying out some basic development work for such assessment as a component of its PAGIS programme (Performance Assessment of Geological Isolation Systems; Cadelli *et al.*, 1984). This uses real sites and fully-fledged repository models, linked with geological, hydrogeological, geochemical and climatic data from the nominated areas. The programme is focused on deep disposal in granites, salt and clays. At the time of writing the process of assembling data and discussing methodologies was underway, but no calculations had been performed. This is clearly a long-term project.

CHAPTER 11

Model Validation and the use of Natural Analogues

Whenever we have discussed the construction of performance models or the assessment of their results, we have frequently dealt with very large-scale and complex interrelated processes, or with extremely long time scales. This is an inevitable consequence of trying to predict behaviour in the geological environment. It raises the question of how confident we can be in the results of the calculations and, at an even more basic level, how confident we are that the processes we are assuming to occur will actually occur at all. Probabilistic models can cater for the last of these questions, but provide no help with the prediction of the effects and consequences of processes.

As a result there has been increasing concern about how we can validate the models which we use in performance assessment. *Validation* should not be confused with *verification*. The latter is simply the process of ensuring that the calculations associated with any particular model actually give the correct numerical answer for the assumptions made, and that they are reproducible by other similar or related models. For some years now the international modelling community have been carrying out detailed intercomparison of models to this very end, and as an aid in developing more sophisticated models. Two major projects have now taken place; INTRACOIN and HYDROCOIN, the first dealing with some general radionuclide transport codes (Larsson *et al.*, 1983), the second specifically with groundwater flow models (SKI, 1985). There is also an international biosphere model comparison underway, called BIOMOVS. This was initiated by the Swedish National Institute of Radiation Protection in 1985. A fourth major programme called INTRAVAL started in 1986 (initiated by the Swedish Nuclear Power Inspectorate—SKI), which will aim not only at comparing coupled transport codes in more detail, but at actually validating them against both *in situ* experiments and what are known as *natural analogues*.

THE CONCEPT OF NATURAL ANALOGUES

Natural analogues are seen by many as the only reliable means of model validation where processes are involved which concern time scales or experimental system sizes/complexities which cannot be replicated reliably in the laboratory. While some aspects of the processes considered in Chapters 5 and 6 can clearly be validated in the laboratory (e.g. simple diffusion coefficients or sorption parameters in simple closed systems), many others cannot. Clear examples of this are the slow alteration of a bentonite buffer material to other clay minerals, the rate of corrosion of metallic containers in the geological environment, the rate at which pore-waters in cement/waste matrices exchange with surrounding groundwaters, and the rate of migration of a radionuclide through complex, mixed rock-formation systems, with differing redox conditions *en route*.

It became clear that the only realistic means of validating models which incorporated such processes was by direct comparison with similar processes, or groups of processes, in the natural environment. These are called natural analogues, although the term is now often used to cover many 'man-made' systems, such as archaeological analogues; the corrosion of ancient metal artefacts for example. Natural analogue studies began by concentrating on the study of radioactive ore deposits, when it was hoped that an analogue for a complete disposal system might be available (i.e a deep ore body might be directly comparable with a deep repository). However, it soon became clear that this was a gross oversimplification which left the validation argument full of holes, and a more appropriate approach was gradually developed, leading to the definition (Chapman *et al.*, 1984) of five 'principles' of the use of natural analogues:

(1) The process involved should be clear cut. Other processes which may have been involved in the geochemical system should be identifiable and amenable to quantitative assessment as well, so that their effects can be 'subtracted'.

(2) The chemical analogy should be good. It is not always possible to study the behaviour of a mineral system, chemical element or isotope identical to that whose behaviour requires assessing. The limitations of this should be fully understood.

(3) The magnitude of the various physico-chemical parameters involved (P, T, pH, redox conditions, concentration, and so on) should be determinable, preferably by independent means and should not differ greatly from those envisaged in the disposal system.

(4) The boundaries of the system should be identifiable (whether it is open or closed, and consequently how much material has been involved in the process being studied).

(5) The time scale of the process must be measurable, since this factor is of the greatest significance (the *raison d'être*) for a natural analogue.

Although analogue studies have been taking place for many years (the earliest and most widely known being the work on the natural reactors in the uranium deposits of Oklo in Gabon; IAEA, 1978), there has been a recent upsurge in interest in many countries and there are now major efforts in hand to apply natural analogues in the manner indicated above: that is, process-specific, with very clear and limited objectives. Such has been the interest that the Commission of the European Communities formed a Natural Analogue Working Group in 1985 to coordinate effort in the field. In its first report (Côme and Chapman, 1986) the NAWG endeavoured to rationalize the use of analogues by identifying four principal means of application.

(1) As natural experiments which replicate a process, or a group of processes, which are being considered in a model. This is probably the most quantitative application of analogues, which allows confident constraints to be placed on, for example, extrapolations of laboratory experiments to larger time or space scales.

(2) For determining the bounds of specific parameter values. This application would be most useful at the stage where a modeller needs limiting values on a parameter, but can obtain these from any or many geological systems. The origins of the data are not particularly important, and need not be linked to the process being modelled. Diverse sources may be used and a statistical approach adopted. An example of this is thermodynamic or kinetic data, which could be obtained from any system.

(3) As simple 'signposts' indicating which phenomena can occur in the system being modelled by reference to a parallel natural system. This is a purely qualititative application which gives 'yes-no' answers, or indicates the 'direction' of long-term processes. It would be the first means of application used when carrying out scoping exercises.

(4) In an empirical sense to integrate the results of many processes at one site, over long time periods. Not all of the processes involved may be evident, nor may the manner in which they have been linked. Only the end result is important, and in this sense this application is the most directly useful to a safety assessment (as distinguished from the individual models which comprise it). An example might be to determine whether there is any surface radiological manifestation of a deeply buried uranium ore body.

Many excellent analogue studies have taken place over the last five years, and rather than trying to summarize them all, we have taken as an example one area where they have been applied to the most important modelling aspects of a particular waste-disposal system. This example is their application to HLW disposal in crystalline rocks. The first part of the study (Chapman *et al.*, 1984; in which the reader will find more details and a full guide to the analogue literature) was an attempt to define which aspects of a complete system performance model were most in need of analogue validation. This could be done by a simple assessment of the sensitivity of the results of the model to the various mechanisms which were included in the performance assessments. The following section outlines how analogue data which were already available at the time of the study could be applied to the Project Gewähr safety analysis considered in detail in previous chapters (NAGRA, 1985). For convenience the discussion is broken down either by process considered or by the particular barrier to which it applies.

NATURAL ANALOGUES OF SPECIFIC BARRIER MATERIALS

Behaviour of borosilicate glass (HLW matrix)

The following parameters are of interest in assessing vitrified waste:

(a) long-term stability (with respect to recrystallization);
(b) kinetics and mechanism of reaction with water;
(c) formation and composition of surface layers;
(d) leach rate of specific elements and their possible immobilization in secondary reaction products.

Volcanic glasses, an obvious potential analogue for waste glasses, arise from very rapid cooling of magma flows or intrusions. Their composition is very variable, ranging from rhyolitic obsidians (with more than 80 per cent SiO_2) to basalt glasses with about 50 per cent SiO_2, which roughly corresponds to the SiO_2-content of waste glasses. However, the Fe- and Al-rich basalt glasses contain practically no boron or lithium and thus have properties different from those of the waste borosilicate glasses.

Nevertheless, basalt glasses show that recrystallization at low temperatures requires around 10^7 years, which is considerably longer than the assumed lifetime of vitrified waste (1.5×10^5 years). Recrystallization is therefore not expected to affect the behaviour of the waste glass within its lifetime.

The hydration rate of natural glasses is basically dependent on the availability of water. Hydrated glass (e.g. perlite) releases elements such as uranium to warm groundwater faster than fresh glass. Cooling and hydration cause stresses which lead to fracturing of larger glass bodies and thus to an increased surface area with correspondingly increased corrosion. Basalt glass surfaces alter rapidly on contact with water. The iron-rich clay layer (palagonite) produced appears to protect the glass for about 10^6 years. On leaching of the glass, uranium and the rare earths (mainly La and Pr, but not Ce) are retained in this surface layer. The layer often contains strongly sorbing zeolites such as chabasite and analcime.

To summarize, natural analogues for the glass matrix indicate that:

(1) Recrystallization (devitrification) is insignificant on time scales relevant to the disposal concept considered.

(2) Natural corrosion mechanisms basically agree with laboratory observations. If only a small amount of water is available, the kinetics of corrosion can be slowed down considerably.

(3) Hydration layers are produced in low water-flow conditions. These act as a crust which can protect the glass from further attack by water for a long time.

(4) Some secondary minerals in the hydration (or transformation) crust show a high sorption capacity for certain elements.

The HLW canister (cast steel)

The model assumption for the canister in the NAGRA safety analysis was a corrosion rate of about 30 μm/year and a minimum lifetime of 10^3 years. The corrosion products then form an efficient redox buffer for more than 10^6 years. With respect to natural analogues, the following issues are of interest:

(a) the kinetics and mechanism of iron corrosion under the expected geochemical conditions;
(b) the long-term stability of the iron corrosion products and their effectiveness as reducing agents.

Investigations of iron meteorites as analogues have, to date, given only qualitative results. They are composed not of iron but of Fe/Ni alloys and their corrosion conditions cannot generally be quantified. Nevertheless, a series of corrosion rates has been estimated and some samples have survived at or near the earth's surface for up to 20 000 years.

The corrosion rates of archaeological finds of iron show a notable consistency. For investigated samples, values lie between 0.1 and 10 μm/year with a few values below 0.01 μm/year. The objects came from a wide range of environments: dry, wet, oxidizing and anoxic. Some Roman iron objects which probably corroded in contact with air in the wet rubble of an old spring showed a corrosion rate of only 1 μm/year.

Natural and archaeological analogues therefore suggest that the corrosion rate of 30 μm/year assumed for the safety model is conservative, especially under the anticipated reducing conditions.

The bentonite backfill in the HLW repository

The most important functions of the bentonite layer surrounding the waste canisters as derived from the near-field analysis are:

(a) forming a barrier with very low permeability (meaning that near-field transport is primarily by diffusion);
(b) preventing transport of radionuclides as particles or colloids;
(c) forming a thermal and mechanical buffer

The most important properties of the bentonite to be investigated by natural analogues are its stability and its effect on the transport of radionuclides.

With regard to stability, it is known that smectite-rich clays can alter to illites (at high temperatures), provided there is a sufficient supply of potassium. This process can also be observed in nature, for example in thick sequences of sediments (e.g. Gulf zone, USA), the age, temperature, history and pore-water chemistry of which are known. Such natural analogues show that illitization is much slower than would be expected on the basis of kinetic models and a regular supply of K^+. Instead of a 100 per cent alteration in 10^5 years, only 0.3 per cent in 10^6 years was observed. Temperatures over 100 °C are expected in the bentonite backfill only for a short time. However, natural hydrothermal systems show that, even at 100–200 °C and with a sufficient supply of K^+, alteration occurs only slowly (in the order of 10^6 years).

Concrete and Cement

Concrete and cement are not only important immobilization matrices for some types of LLW/ILW but are also widely used as structural and sealing materials. Analogue studies should give information on their mechanical long-term stability and leaching behaviour in relevant chemical environments.

There are two types of historic and archaeological analogues: industrial constructions (*ca.* 100 years old) with concrete of relevant composition, and mortars and cements up to 3000 years old, with widely varying chemistry. In the first group there are many examples of dams, piers and canals which have remained intact for more than 50 years. The second group is more variable but also contains hydraulic installations which have been very stable under relatively aggressive conditions. There are, however, some cases where rapid destruction of concrete was observed, either in aggressive chemical environments or through intensive microbiological activity.

A valid conclusion from these findings would be that concrete structures can survive for 10^3 years in surface environments but it is not yet possible to draw quantitative conclusions for particular corrosion or leaching conditions.

Bitumen

Bitumen may be used in the LLW/ILW repository as a solidification matrix for certain waste types. Geological deposits of tar and pitch can serve as natural analogues. These show good stability under very unfavourable conditions (e.g. on the earth's surface) which allows an expected lifetime of 10^5 to 10^6 years to be deduced. However, this can be drastically shortened by microbiological activity. Archaeological samples containing tar or pitch have proved to be very stable over *ca.* 500 years. There are, however, many examples to the contrary where bituminous covering material was rapidly destroyed in damp soil, possibly again with involvement of micro-organisms. In the absence of microbiological attack, therefore, bitumen can be regarded as very stable.

NATURAL ANALOGUES OF RADIONUCLIDE RELEASE AND MIGRATION

Natural analogues of radionuclide release and migration have been extensively studied in the last few years. Two of the best known 'global' analogues—Oklo and Morro de Ferro—will be briefly described as illustrations of the general approach. In order to validate models used in the safety assessment chain, more specific analogues are, however, required and these are considered for:

(a) solubility limits and speciation;
(b) retardation during groundwater transport;
(c) redox conditions.

Oklo and Morro de Ferro

The best known example of a natural analogue is the Oklo uranium ore deposit in Gabon, Africa (Brookins, 1976). About 2000 million years ago, spontaneous chain reactions occurred in several identified 'reactor zones' and these continued intermittently for around 10^5–10^6 years (Walton and Cowan, 1975). About 1000–2000 tonnes uranium were present as 'fuel', of which around 6–12 tonnes of ^{235}U underwent fission, and about 4 tonnes of plutonium was produced.

The natural reactor operated at temperatures of 400–600 °C and at pressures of 800–1000 bar, which corresponds to a depth of about 3.5 km. The uranium isotope ratios used as a measure of the 'burn-up' of the nuclear fuel are normally constant within small rock specimens and ore grains but characteristically different for different reactor zones, which indicates that no (or little) uranium was transported between the reactors. However, there is some indication of migration of uranium with anomalous isotopic composition into the surrounding altered rock for a distance of a few metres. Radioactive fission products decayed within the geological time periods involved but have left characteristic fingerprints in their stable daughters. The isotopic composition of the latter can be determined by mass spectroscopy and gives information on the extent of migration from the reactor zones into the surrounding rock during or after the reaction period.

In recent years, Morro de Ferro in Brazil (Eisenbud et al., 1984) has also been investigated. This body of iron ore with an estimated 30 000 tonnes of associated thorium and 50 000 tonnes of rare earths (mainly neodymium and lanthanum) lies near the surface on a hill of strongly weathered and hydrothermally altered rock which is exposed to erosion and leaching by rain-water. The Th content in the out-flowing groundwater is however only ca. 0.05 g/l and that of La 0.28 g/l. From simple calculations, mobilization rates of 1.4×10^{-9}/year for La result. Data recently measured for ^{228}Ra gave 10^{-7}/year. The total uranium content of the ore deposit is only ca. 100 tonnes. It is postulated that this figure was very much higher at the time of origin of the ore body some 60 million years ago, and that the U was washed out under the prevailing oxidizing conditions. Transfer factors for Th, Ra and the rare earths through the food chains to man and animals were also determined around Morro de Ferro. Th- and Ra-concentrations in plants, vegetables and milk are, at most, 10 times higher in the vicinity of Morro de Ferro than in similar products from typical agricultural areas in the northern hemisphere.

Solubility limits and speciation

In the NAGRA Project Gewähr safety analyses, actinide solubility and speciation are determined by use of thermodynamic equilibrium models. These are checked by comparison of the calculated concentrations of selected elements (e.g. U, Th) with those found in natural groundwaters. The natural release of these elements from the rock is an analogue for release from the waste matrix. Although there are many summaries of water analyses in relevant granite formations, these are generally not extensive enough to allow speciation to be examined in a dependable manner. Inadequacies in sampling and lack of equilibrium are further sources of errors. Despite all this, data can be obtained from such measurements and provide important checks on the models used (Schweingruber, 1983).

Use of analytical data as an analogue for speciation has only been attempted in a few cases as, for example, the Th- and La- content in water from Morro de Ferro considered above.

The following conclusions can be drawn:

(1) Extensive analysis of groundwater from the host formation allows the rock itself to be used as a source of certain relevant elements (analogue of the HLW matrix). Assuming solution equilibria, such data can be very useful in checking thermodynamic models.

(2) Direct extrapolation of speciation data from near-surface soil conditions to those expected in deep groundwater is inapplicable for many radionuclides. Such data are, however, useful for checking speciation models, particularly in the case of elements with uncertain or non-existent thermodynamic data.

Retardation of radionuclides during transport

During groundwater transport, there is interaction between the dissolved radionuclides and the mineral surfaces present. Separation of the particular mechanisms involved is very difficult in natural systems but, in general, they result in retardation of the dissolved species in relation to groundwater (or in relation to an inert species in the case of diffusive transport). This retardation may, for example, be due to sorption of the solute onto mineral surfaces which, in turn, is dependent in a complex manner on many parameters such as speciation, surface structure, temperature and so on.

The following parameters could be investigated by analogue studies.

Retardation

There are many analogues for radionuclide migration over a wide time range. Preconditions for use of such an analogue system are:

(a) spatially and temporally defined source of the elements of interest;

(b) a well-defined hydrogeological situation during migration;
(c) a measurable concentration profile resulting from migration.

Examples of such analogue systems, are: the Oklo natural reactor; ore deposits; aquifer redox fronts; discontinuities in sediments; fallout from nuclear weapons tests.

The following conclusions result from studies of these systems:

(1) Investigations of the Oklo reactor give only guiding qualitative data. High-temperature processes (e.g. in connection with magmatic intrusions) are not relevant for most HLW disposal concepts.
(2) Investigations of the hydrology in the vicinity of ore deposits give much qualitative data on site-specific radionuclide migration which in turn contribute to quantitative development of relevant models. Weathering profiles and aquifer redox fronts can be investagated by isotopes of the U and Th decay series, whereby the extent and the kinetics of radionuclide retardation can be derived.
(3) Sediment profiles which have dateable geochemical anomalies have potential as a means of quantifying retardation during diffusion of solutes through clay strata.

Sorption:

Sorption can be calculated for equilibrium groundwater systems from the exchangeable radionuclide concentration on the solid and in the liquid phase. The greater the sorption, the greater the retardation. Analogue studies often give data for both sorption and total retardation. Such data on natural sorption (or retardation) were taken into account when specifying the database for the Project Gewähr transport models (McKinley and Hadermann, 1984).

Matrix diffusion:

Rock profiles from a disturbed zone through which water with a sufficiently high uranium content has flowed can be used for matrix diffusion analogue studies. Interpretation is made difficult by a usually blurred transition from the strongly permeable alteration layer to the micro-fissures in the unaltered rock. For example, during a hydrothermal phase (100–800 °C), the penetration depth of uranium enrichment may reach about 3 cm and the subsequent alterations at lower temperatures (< 100 °C) may remain within this distance. Unfortunately, these and other results are not sufficient to allow unambiguous assessment of the extent of matrix diffusion in relevant situations.

Redox conditions

A potentially important factor in near-field chemistry is the radiolysis of water. Because of the high radiation levels required over long time periods, the selection of natural analogue systems for the effects of radiolysis has been limited to the Oklo natural reactor. The radiolytic effects were investigated by Curtis and Gancarz (1983) with the following results:

(1) The radiolytically formed H_2 did not behave inertly but diffused into the structure of the clay minerals and reduced Fe^{3+} to Fe^{2+}. This shows that a model with an exclusively oxidizing radiolysis effect can be considered as conservative.
(2) Although the Fe^{2+}/Fe^{3+} ratio in the reactor zones points to reducing conditions, the actual redox conditions are not known and apparently vary locally.
(3) Considering the complex nature of the redox conditions in the near-field, the simple assumption of an oxidation front moving through the rock does not appear to be justified in the case of Oklo.

In the NAGRA concept (see Chapter 5), a possible oxidation front would be hindered, by canister corrosion products, from spreading beyond the bentonite barrier for at least 10^6 years. Some uncertainties could be clarified by natural analogues, for example:

(a) the mechanism by which natural redox fronts are produced and the behaviour of redox-sensitive species at such a front;
(b) the extent to which Fe(II) in minerals (primarily in intact granite) is available as a redox buffer for aqueous systems and the kinetics of such buffering;
(c) the effectiveness of corrosion products as a redox buffer.

FUTURE OF ANALOGUE STUDIES

From the above example it can be seen that many questions of detail require some form of analogue validation. These will vary considerably from one waste-disposal concept to another, in particular in terms of the degree of significance which particular processes attain in different system models. At the time of writing a number of completely new analogue studies were commencing which addressed several of the means of application mentioned earlier, from the empirical (e.g. studies of the 400 m deep Cigar Lake uranium deposit in Canada, which had not been detected by standard surface geochemical prospecting techniques, and hence had no surface manifestation, or 'signature', with consequent implications for long-range radionuclide transport), to the highly process-specific (e.g. studies of chemical speciation and microbiological activity in highly alkaline springs, replicating cement pore-water chemistry).

There is clearly considerable scope yet in the use of natural analogues in model validation, but it must be stressed that they are by no means a 'modeller's panacea' in terms of answering all the problems of confidence levels. In particular, individual analogue studies must not be overinterpreted, as has sometimes happened in the past. The natural analogue probably represents the only widely understood means of demonstrating confidence, and more effort needs to be put into enhancing its credibility.

CHAPTER 12

What Does it all Mean?

Throughout this book we have endeavoured to treat nuclear waste management and disposal as a reasonably standard, albeit very complex, scientific and technical issue. Geological disposal strategies can be designed making use of available technology and, by the well-established methodology of mathematical modelling, they appear to satisfy defined performance objectives (usually set in terms of acceptable risks or radiation doses). In the previous chapter we dealt with techniques for answering the principal uncertainty involved in this approach; how do we justify our confidence in the very long term predictions which are produced by these performance-model calculations?

If significant doubts remain after these attempts at model validation, then the designer of a waste facility can always attempt to overcome them by demonstrable overdesign of the disposal system. In practice there is always a substantial margin of safety incorporated in any reasonable engineering design, whether it is a bridge girder or a car-engine part. In waste management terms this is clearly illustrated by the systems currently being advanced for HLW disposal. The results of the Swiss and Swedish safety analyses produced radiation doses which, even under the most pessimistic of likely scenarios, would be totally lost in the natural background. The engineered barrier systems suggested are enormously expensive in terms of material alone. For example, the Swedish proposals would use up to 72 000 tonnes of copper and 47 000 tonnes of lead to dispose of only about 6000 tonnes of spent fuel. Clearly this approach is acceptable only because such small volumes of waste are involved. In most European countries the HLW arisings to the end of the century, whether as spent fuel or in reprocessed and conditioned form, generally amount to only a few thousand tonnes. This material also represents more than 95 per cent of the long-lived waste radioactivity arising from substantial nuclear power programmes operating over several decades, so the concept of investing a lot of money in overdesign of a disposal facility does not seem too much out of place.

This approach works less well as one considers the progressively larger volumes of waste which arise as one moves into lower and lower activity ranges.

In the UK, for example, the volumes of waste increase by more than an order of magnitude for each step from HLW–ILW–LLW and the wastes become more diverse and difficult to treat in a simple uniform manner, such as is applicable to HLW. At the same time the potential hazards of the waste become less and less. In these circumstances it is clearly less acceptable to produce a monstrously overdesigned disposal system, as the costs rapidly become out of all proportion to the problem. Consequently, there is a move towards adopting some rational means of cost-benefit analysis to the design of LLW/ILW disposal systems, where benefit is calculated in terms of realistically achievable dose or risk reductions (both from disposal operations and from the long-term behaviour of the repository), based on the ALARA principle. Once confidence is established in the use of this technique, it may be applied to the HLW designs mentioned above. It should be remembered, in any case, that these two HLW disposal concepts were intended only to show that the task could be carried out safely and acceptably, using available technology. One might now expect the groups concerned to step back and produce more economic, but nonetheless acceptable solutions.

All of this brings us back to the results of the safety assessments we considered in Chapter 10. In our Preface we rather flippantly observed that waste disposal had been a problem ever since Ug the Caveman first wondered where to throw the accumulated half-chewed bones from the floor of his cave. Although he may have known nothing of disease and sanitary practices, the consequences in terms of unpleasant smells and flies must have been obvious. In much the same manner, the immediate impacts of present·day 'environmental nasties', such as an abbatoir or chicken farm opening up next door, are very obvious, and inevitably lead to considerable and vociferous opposition. This has become known as the NIMBY syndrome (Not In **My** Backyard), and is an inevitable response faced by organizations trying to manage radioactive wastes, especially when seeking new disposal facilities. To many people the prospect of a neighbourhood 'radwaste dump' must seem infinitely more unpleasant than most other 'developments', especially if imagined in those emotive terms. We hope that by going through all the issues involved in designing one of these 'dumps' we will, if nothing else, have demonstrated that they are considerably more sophisticated than the smelly landfill sites in which our domestic garbage ends up.

This comforting thought does not, however, answer the basic question as to what those very small calculated risks and doses in Chapter 10 actually signify with respect to everyday life; we have after all seen some vanishingly small numbers and some enormously long time spans referred to earlier on. What do these values mean?

Safety analyses provide three types of information:

(a) dose levels which measure the amount of harm caused;
(b) time scales which indicate when that harm will arise;
(c) probabilities which indicate how likely this is to occur.

In addition there is also a further uncertainty (usually not explicitly stated) as to how accurate each of these values are. We can now consider these point in turn.

The most obvious and easily assimilable way of assessing doses is by comparison to those arising from natural background radiation. As considered in Chapter 1, natural dose rates vary widely and the limits set for those which might result from repository releases are less than would be experienced by an individual moving to a region of higher altitude, or to an area with more radioactive bedrock, or having occasional medical X-rays—none of which are normally considered to be hazardous, despite being radiologically 'harmful'. Needless to say, the 'harmfulness' is so low that many thousands or millions of people would have to be considered before any statistically significant effects became noticeable, but the notional risk is still there. In general terms it is assumed that as radiation exposures decrease below natural background values the harmfulness decreases proportionally, but at certain levels it becomes completely unmeasurable. The realm of statistics is, without doubt, arcane when trying to extract any social significance out of very low notional risks to large hypothetical populations. The statistical risk (associated with a radiation exposure of 1 mSv each year) of 'one death per year in a population of 100 000', which is the current UK objective for exposure to all doses other than natural and medical exposures, is difficult to comprehend if extrapolated to assess the KBS-3 doses of less than one-thousandth of this value. Does this correspond to a risk of one death each year in a population of 100 million? Or does it mean one death every 100 thousand years to a population of 1000 living on top of a repository? The natural background is itself 1000 times higher than predicted additional doses from the waste, at about 1 mSv. Does the natural background also result in one death per year in each group of 100 000 people? No matter which way one looks at the numbers, they seem very small when equated with the risk of smoking, or even of crossing the road.

The second factor is the time-scale of releases, which extends beyond common understanding. The safety analyses usually complicate this problem by use of logarithmic time-scales, which are convenient for modellers but tend not to show things in comprehensible perspective. This can be illustrated in Fig. 12.1 which compares the timescale of human evolution with the repository time-scales considered here.

A pragmatic approach to time-scales would be that if you can really imagine what it would be like at that time, it is meaningful—if not, forget it. Computer models will happily predict releases after the expected lifetime of the earth. This is one reason why some groups have suggested application of a cut-off time beyond which doses should not be calculated. A figure of 10 000 years has been suggested but this is debatable. To the sociologist or food scientist it is probably one or two orders of magnitude too long for rational prediction; to the climatologist in northern Europe it may be about right as the time to the next ice age; to geologists who are used to playing with big numbers and confident of their ability to predict global evolution over many millions of years, it is probably too short. It is unlikely that most people are too concerned about events after the next ice age.

228

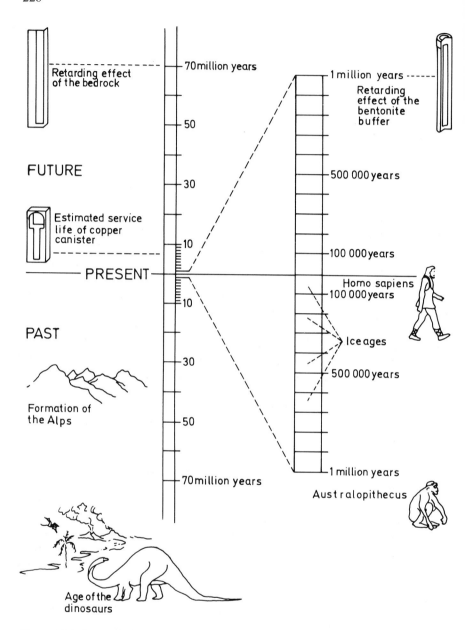

Figure 12.1 Comparison between time spans for the durability of barriers and the same time spans backwards into the past (after KBS, 1978)

Finally we must take into account the uncertainties due to limitations of knowledge, lack of data and the sheer complexity of natural systems. There are only a few individual parameters in the safety assessment model that could be proven to a critical scientific review to be accurate, even to within about an order of magnitude, in all possible circumstances. Indeed, measurements of some may show irreducible uncertainty ranges of several (or even many) orders of magnitude. In safety assessment this uncertainty is generally taken into account by selection of conservative values which would lead to overprediction of doses. It is assumed that, with this approach, even if one or two values may not be truly conservative, the overall result of the multiple barrier assessment will be. Natural analogues can also be used to check that the final results of analyses are reasonable.

Whether we like it or not, radioactive wastes are now part of our lives. If all nuclear programmes, both civil and military, were to stop today, there would still remain a vast quantity of wastes to deal with. They are very much a problem of today, a legacy that future generations might not thank us for passing on.

In this book we have tried to demonstrate that they can be disposed of in a proper fashion, with convincing margins of safety, and with due regard to the actual magnitude of the dangers involved. In the past they may not always have been treated in this way, and we can offer no reassurance that they will always be handled responsibly in the future. We merely state that, from the viewpoint of our own disciplines of earth sciences, physics and chemistry, the problem appears quite tractable. Radioactive wastes require very similar treatment and disposal techniques to other toxic wastes. At present there is considerably more effort going into solving the problems of the former, and scant attention being paid to the potentially far more widespread and potent problem of chemical wastes. This is rather at odds with the comparative scales of the two problems, and must appear rather anomalous to anyone concerned with the overall quality of the environment.

We would like to end this book by sticking our necks out a little. By far the greatest public concern is voiced over the issue of HLW management, and very sophisticated deep repositories have been designed to ensure the safe disposal of what are in fact very small volumes of material. All plans to date, at least in the western industrial countries, assume that such waste must be handled on a national basis, and even if fuel is reprocessed abroad, solidified HLW will be returned to the country of origin. As the cost and environmental impact of one repository to cope with, say, all European waste, would be much less than those of many smaller disposal facilities, there are obviously considerable incentives for handling HLW on an international basis. If such an approach were politically feasible, it would not only offer cost-benefit and environmental gains, but might also lend itself to management by a recognized international organization, a factor which might also increase public confidence.

The topic of nuclear waste tends to polarize debate; many people within the nuclear industry consider that the technical problems are in themselves minor, and much of the current research effort is unnecessary. On the other hand, some environmental groups regard nuclear waste as an insoluble problem, an ultimate evil with which mankind is saddled. Some believe that this gulf is unbridgeable, regardless of the amount of information available or the degree of communication which takes place. Ultimately, we believe, this disparity will be resolved, and we hope that this book has in some way provided a middle ground which is useful to both sides of the debate.

APPENDIX I

International Organizations and National Research Programmes

INTERNATIONAL ORGANIZATIONS*

IAEA

The International Atomic Energy Agency (IAEA) is part of the United Nations family and, with 111 members, is the largest organization of this type. The IAEA is concerned with all peaceful aspects of nuclear energy and its main functions are dissemination of information, coordination and promotion of technological developments, and provision of safety guidelines.

The IAEA waste management programme is divided into three categories:

(a)   Waste handling and treatment;
(b)   Underground disposal;
(c)   Environmental aspects of nuclear energy.

The programme consists primarily of the organization of international symposia and the preparation and publishing of reports. The reports are written by advisory groups and consultants from the various member states. Their preparation provides an opportunity for experts to meet and discuss, as well as a means of disseminating information and ensuring consistency of approaches to waste management. The IAEA also conducts Coordinated Research Programmes, in which various institutes who are engaged in work in a particular field exchange results and coordinate their work. For high-level waste the only disposal option currently being considered by IAEA is geological disposal on land.

---

* Much of the content of this section was derived from the excellent review of Parker *et al.* (1984) which provides more detailed information and background references.

NEA

The Nuclear Energy Agency (NEA) is a branch of the Organization for Economic Cooperation and Development (OECD) whose membership includes most of the highly industrialized non-communist countries. Its aims are the stimulation and co-ordination of efforts in its member countries in various aspects of the utilization of atomic power.

The objectives of the NEA programme on waste disposal are to promote:

(a) exchange of information between member countries;
(b) co-ordination of national research and development activities;
(c) international co-operation, through co-operative experiments and research projects.

The first two objectives are achieved by organizing workshops on technical topics, sponsoring symposia, and through regular meetings of committees and expert groups, in particular the Radioactive Waste Management Committee, the Coordinating Group on Geological Disposal, and the Seabed Working Group. The third objective involves organizing specific multinational projects, such as the Stripa mine project (see Chapter 7) and the International Sorption Information Retrieval System (ISIRS) maintaining other chemical and nuclear physics databases, and promoting continuous cooperation, for example joint research vessel cruises to ocean areas of interest for high-level waste disposal. NEA differs from IAEA in that its programme covers both geological and ocean disposal of high level waste.

In 1986 the Coordinating Group on Geological Disposal was replaced by the Advisory Group on *In Situ* Research and Investigations for Geological Disposal, whose main function will be to coordinate effort and facilitate information exchange in the fields of performance assessment and underground research laboratories.

ICRP/UNSCEAR

The International Commission on Radiological Protection (ICRP) is a non-governmental organisation which issues recommendations on radiation protection which form the basis for dose limitation regulations by many countries and by other international bodies (e.g. IAEA, NEA). ICRP has a close relationship with the United Nation's Scientific Committee on the Effects of Atomic Radiation (UNSCEAR), which is a standing committee of the United Nation's General Assembly. The main task of UNSCEAR is to assess the radiological impact of all radiation sources, both natural and artificial. The reports of ICRP and UNSCEAR together are generally considered to be the most authoritative international sources of information for radiation risk estimates.

CEC

The mandate of the programme on radioactive waste management of the

Commission of the European Communities (CEC) is to coordinate research activities between its member states. Unlike bodies such as IAEA and NEA, the CEC provides considerable funding for waste disposal research. Its programme has two components: direct action, consisting of work carried out at the Joint Research Centre at Ispra in Italy, and indirect action (consisting of partial funding of research activities in the EC countries). The first five-year CEC waste management research programme (1975–79) included many activities in the field of geological disposal of high-level waste.

In the second five-year programme (1980–84) the research was extended to include subseabed disposal, as well as geological disposal, and initiated major coordinated research programmes such as MIRAGE (Migration of Radionuclides in the Geosphere: Côme, 1985) and PAGIS (see Chapter 10). The results of both programmes were published in various technical reports and presented at two international conferences (Simon and Orlowski, 1980; Simon, 1985).

The third five year programme began in late 1985 and is in many ways an upgrading and continuation of the second programme, putting more emphasis on URL activities, as well as retaining PAGIS and MIRAGE. Both the direct and indirect action programmes cover central issues in shallow and deep disposal, including geochemical studies, properties of waste-forms and engineered barriers, repository construction techniques, mathematical modelling and risk analysis.

CMEA

The Council for Mutual Economic Assurance (CMEA) coordinates research in the USSR and most of the Eastern European countries. Since this organization does not publish details of its programme on a regular basis it is difficult to describe its activities. It is known that CMEA is considering only the geological disposal option, but beyond this very little information is readily available.

MAJOR NATIONAL PROGRAMMES ON DISPOSAL OF LONG-LIVED WASTES

Argentina

In the late 1970s Argentina began a programme of research into various methods of disposing of HLW from the country's power reactor programme. These will be vitrified wastes, generated by a reprocessing plant which is in an advanced state of construction. At present about 3000 containers of waste are expected to be produced, and a repository is planned to be constructed in granitic rocks. Of seven areas initially examined, the most promising appeared to be at Sierra del Medio (Chubut province), and detailed site investigations are underway.

Belgium

The Belgian programme is focused on disposal in clay formations, and in particular on the Boom clay which underlies the Mol nuclear research site. At present it is assumed that spent fuel will be reprocessed abroad and vitrified waste will be returned for disposal. The main activities consist of investigation of the physical, chemical and hydrogeological properties of the Boom clay (via *in situ* and laboratory experiments) and safety assessment studies (in cooperation with the CEC Joint Research Centre at Ispra). Work has developed to the stage of constructing a trial shaft and an experimental underground laboratory in the clay at a depth of over 200 m. This provides information on the techniques which will be required in construction of a mined repository in the plastic clays, as well as data for safety-assessment purposes.

Canada

Canada is carrying out research into disposal of high-level waste in the crystalline rocks of the Canadian shield. The main emphasis is on direct disposal of spent fuel with reprocessing currently a low priority alternative. The programme covers all aspects of geological disposal (site investigations, waste form studies, engineering studies, safety assessments and so on). At present it is proposed that eventual waste disposal will take place in Ontario, the only province with operating power reactors. A major programme is underway to construct an underground research laboratory (URL) about 260 m beneath the Whiteshell research establishment in Manitoba to investigate the properties and behaviour of shield rocks.

Denmark

A major feasibility study for disposal of HLW into deep boreholes drilled into a salt dome was completed in 1982 by ELSAM/ELKRAFT. These boreholes would be drilled to about 2500 m, being the deepest disposal concept so far advanced as a national programme.

Federal Republic of Germany (FRG)

Research in the FRG is primarily concerned with deep salt formations, the majority of the effort being devoted to investigating the Gorleben salt dome and to experimental work at the Asse salt mine. The German programme for HLW assumes reprocessing and vitrification of wastes. This programme, called PSE (Project Safety Entsorgung, where Entsorgung is a term incorporating all aspects of radioactive waste management), aims to develop Gorleben into an operating disposal facility by 1999, construction commencing in 1995. In 1986 the sinking of an exploratory shaft was underway, preliminary site exploration by drilling having been almost completed.

The Konrad iron ore mine, near Salzgitter, was also being developed for use as a repository for non-heat emitting wastes. A licence for construction and operation was expected by 1987, with operation to begin by 1989. The Asse salt mine, extensively used for experimental and trial disposal purposes, might also be available for development as a repository should a future need arise.

Finland

The Finnish research and development programme closely follows that of Sweden, since both involve disposal in the hard crystalline rocks of the Scandinavian shield. Finland operates two nuclear power plant sites, Loviisa and Olkiluoto. Both sites had been investigated in detail by 1985, and plans for repositories prepared. These facilities will be used for deep disposal of LLW/ILW, as well as for future decommissioning wastes. They are very similar in concept to the Swedish SFR facility. Construction will commence in 1988, and operation in 1992.

Field investigations for a disposal site for spent fuel begin in 1986, after three years' work screening the whole of Finland in which 101 potential sites were located. Up to 10 sites will have been investigated by 1992, and one selected in 2000. A repository is expected to be operational by 2020. At present spent fuel from the Loviisa plant is returned to the USSR, and an interim wet store is to be commissioned for the Olkiluoto plant in 1987.

France

The French programme is similar to the Canadian one in that it initially focused on granite formations and, to a lesser extent, on subseabed disposal. Recently, as a result of the report of the Castaing Commission (Castaing, 1983) consideration has been given to other (sedimentary) rocks, although the emphasis on crystalline rocks remains. A deep, purpose built, research cavern is planned, and existing mines are already in use to perform hydrogeological testing (e.g. at Fanay-Augéres). As it offers a commercial reprocessing and vitrification service to other countries, this is the primary disposal route in France for HLW. The Commissariat de l'Energie Atomique (CEA) operates a shallow burial site for LLW/ILW at Centre Manche (Cap de la Hague, Bretagne), through the official waste disposal organization, ANDRA. Since this site will be filled during the 1990s, two additional near-surface disposal sites were being sought in the mid-1980s, one of which had been designated for Soulaines (near Troyes) in 1985. This site is anticipated to have an operational life of about 30 years.

Italy

Present plans favour the development of a deep borehole disposal facility for vitrified HLW and cladding hull wastes (ILW) in Plio-Pleistocene plastic clays, rather than using a mined repository. An underground research laboratory is being constructed in clays in Sicily and it is proposed to carry out a full scale demonstration of deep borehole disposal technology soon.

Japan

Both geological and subseabed disposal are being investigated in Japan. Japan is committed to reprocessing with solidification of resulting wastes in borosilicate glass or possibly SYNROC. The rock types being studied include granite, and mixed mudstone-shale-sandstone sequences. Research includes field surveys, laboratory experiments and development of mathematical models. *In situ* tests are being developed in existing mines, and a near-surface laboratory has been constructed in granite. Several cruises have taken place to survey potentially suitable subseabed disposal sites.

Japan is a participant in several international projects, such as Stripa, the Seabed Working Group and a joint programme with Switzerland and Sweden on the leaching of fully active vitrified waste.

Netherlands

The options being studied in the Netherlands are disposal in salt domes on land, disposal in salt domes under coastal seas and subseabed disposal in the deep ocean. The salt-dome research has so far been confined to generic studies. Deep-ocean research has concentrated on site surveys in the Atlantic.

Sweden

The Swedish programme is one of the most advanced in the world, concentrating on disposal of all types of waste in crystalline rocks of the Scandinavian shield, at various depths. The Swedish Nuclear Fuel Supply Company (SKB, until 1984 known as KBS) operates a major research and development programme, with other organizations such as SKI (Nuclear Power Inspectorate), SKN (National Board for Spent Fuel) and SSI (National Institute for Radiological Protection) also sponsoring research and acting in advisory or regulatory roles. The well-known Stripa underground research laboratory is situated in central Sweden. Site investigations are underway with a view to selecting two or three sites for detailed assessment for a deep repository for spent fuel around 1990. Application for a development licence would occur around 2000, with construction beginning in 2010. Meanwhile, an underground store (about 30 m deep) for 3000 tonnes of spent fuel (the CLAB facility) has been constructed at a coastal site (Oskarshamn). LLW and ILW are to be disposed of in a repository (the SFR facility) currently under construction at Forsmark some hundreds of metres offshore, at a depth of 50 m below the seabed, with access by inclined tunnel from the coast. This will be operational by 1988.

Switzerland

The Swiss programme advanced rapidly during the early 1980s to the production in early 1985 of a very detailed proposal for deep disposal of all waste types. The proposal (Project Gewähr) included the results of considerable experimental work and deep borehole drilling in northern Switzerland. The concept suggests disposal of HLW at depths up to 1500 m in crystalline rocks underlying sediments in the north of the country, and the construction of mined facilities in the Alps for LLW/ILW in marl (Oberbauenstock), anhydrite (Bois de la Glaivaz) or crystalline rocks (Piz Pian Gran). An underground rock laboratory is operating in crystalline rocks at Grimsel, and further laboratories are planned in other rock types.

USSR and Eastern Block Countries

Very little up-to-date information is available from these countries, although it is believed that a wide variety of disposal techniques are either under consideration, or have been tried out in the past. A considerable amount of research and development on waste conditioning (vitrification, bitumenization, etc.) is also taking place.

In the USSR, deep disposal of HLW in shafts lined with stainless steel and surrounded by reinforced concrete was described in 1976. Shallow disposal of shorter lived wastes is known to take place in the USSR, and in 1985 there were reports of concrete bunkers being used to store (presumably higher level) wastes from nuclear submarines in Estonia. Liquid wastes have certainly been injected deep into the ground in the USSR (e.g. at Dimitrovgrad), and in 1977 there were unconfirmed reports of disastrous contamination of surface waters as a result of an 'explosion' in reactor wastes in 1957/8 near Kyshtyn in the Urals. In 1979 the USSR was developing intermediate depth injection for some long-lived liquid waste types, in boreholes from 350–500 m and 1500 m deep. This work was linked with a detailed site monitoring programme. The recent reactor accident, at Chernobyl in the Ukraine, has produced large volumes of contaminated materials and topsoil from the reactor site and surrounding area, which present a massive disposal problem. At present, near surface disposal seems the most likely solution for most of the material involved, although radionuclide contents of some types are sufficiently high that some form of engineered barriers will probably be needed.

In 1982 Poland was assessing the suitability of layered Permian rock-salt formations in the north and east of the country for long-lived waste disposal.

UK

The system of waste management in the UK is described in detail in Appendix II.

USA

The USA has a very large and diverse programme of research into nuclear waste disposal. Current policy on HLW is defined by the Nuclear Waste Policy Act of 1982 which resulted in three sites being nominated in 1985 as potential repository locations. These are Deaf Smith County in Texas (salt), Yucca Mountain in Nevada (tuff) and Hanford Reservation in Washington (basalt). After detailed site investigation, one of these three locations will be selected in 1991 with the aim of commencing repository operation in 1998. The US Department of Energy has also nominated a further 12 areas for examination as potential sites for a second HLW repository. These repositories will accept both military and commercial wastes—the former expected to be mainly reprocessing waste (vitrified) and the latter to be spent fuel, of which about 50 000 tonnes is expected to have accumulated by the end of the century.

Military ILW (referred to as transuranic waste, or TRU, in the USA) is handled separately from civilian waste and is intended to be emplaced at the WIPP (Waste Isolation Pilot Plant) facility in New Mexico. Construction of the WIPP should be completed in 1986 and a five year demonstration period involving emplacement of TRU should commence in 1988. Military LLW is disposed of by shallow land burial, grout injection or emplacement in deep shafts on military sites (Savannah River Plant, South Carolina; Hanford Reservation, Washington; Idaho National Engineering Laboratory, Idaho).

Commercial LLW (some 90 000 m^3 per year) is subject to the Low-Level Radioactive Waste Policy Act of 1980, which places the responsibility for ensuring adequate disposal capacity with individual states. This is done on a regional basis, with agreements between neighbouring states, although at present only three sites are operational; Barnwell, South Carolina; Beatty, Nevada; and Hanford, Washington. The search is in hand for further new sites, to ensure better regional coverage, and there is extensive debate on site selection methodology (e.g. Rogers, 1986).

APPENDIX II

Radioactive Waste Management in the United Kingdom

The responsibilities for management and disposal of United Kingdom radioactive wastes were revised in a government White Paper (Cmnd 8607) in July 1982. The disposal of radioactive wastes requires an authorization under the Radioactive Substances Act of 1960, from the relevant government department. This will also apply to the development of any new facilities. Generally speaking, the authorizing body is the Department of the Environment (acting together with the Ministry of Agriculture, Fisheries and Foods) and its equivalents, the Scottish and Welsh Offices.

The Department of the Environment (DoE) also has responsibility for coordinating research and establishing waste-management criteria, and has recently published principles for the disposal of LLW/ILW (DoE, 1984). This responsibility for waste-disposal policy was originally established in a government White Paper in 1977 (Cmnd 6820). DoE contracts research to numerous government and private laboratories, and ensures cooperation in the CEC, OECD/NEA and IAEA research programmes. DOE is advised by an independent group, the Radioactive Waste Management Advisory Committee (RAWMAC), as to whether research goals and results, and guidelines and practices are acceptable. The annual reports of RAWMAC are frequently used to define government policy.

Executive responsibility for LLW/ILW management and disposal has been placed with UK NIREX Ltd, (the Nuclear Industries Radioactive Waste Management Executive). The NIREX directorate comprises staff from the major waste producers, British Nuclear Fuels plc (BNFL), the electricity generating boards (CEGB and SSEB), and the United Kingdom Atomic Energy Authority (UKAEA). NIREX is funded by these groups, as any disposal facilities will be, on a similar basis to the rather emotively named 'polluter pays' principle which is applied to industrial waste management.

Nuclear sites are subject to licensing under the Nuclear Installations Act of 1965 and to periodic checks by the Nuclear Installations Inspectorate (NII); the Health and Safety Executive (HSE) also has a statutory role in securing the safety and welfare of staff. Thus, while many organizations are involved in the UK programme, the outline can be summarized as follows.

DoE has primary responsibility for ensuring that waste-management practices are adequately researched, subject to international codes of practice as applied to UK circumstances, and form part of long-term government policy. NIREX is responsible for designing and implementing the 'hardware' of disposal practices. NII and/or DoE ensure compliance of facilities with UK legislation on radioactive materials. RAWMAC takes an independent role in continuous assessment of the position and in advising DoE in regular meetings and annual reports (RAWMAC, 1980, onwards).

At the time of writing, NIREX were responsible for the NEA supervised sea dump (although this was suspended subject to review in 1983), were investigating sites for near-surface disposal of LLW/ILW, and were looking at options for deep geological disposal of ILW both on land and offshore. As regards existing disposal sites, LLW disposal continued at Drigg in Cumbria, operated by and for BNFL, in addition accepting wastes from the National Disposal Service (NDS). The NDS is operated by UKAEA to cope with wastes arising outwith the nuclear industry (e.g. in hospitals, research laboratories, industry and the universities).

A group of potential sites (Bradwell, Essex; Elstow, Bedfordshire; Fulbeck, Lincolnshire; and Killingholme, Humberside) for a new near-surface disposal facility to handle the large quantities of low-level power reactor operating wastes currently in store had been identified in early 1986, and site investigation programmes began later that year, under the provisions of a Special Development Order. A possible site for deep disposal of long-lived ILW, the Billingham anhydrite mine, never got past the desk study stage, owing largely to problems of ownership and strong local opposition. Early in 1986 the government decided that *all* types of ILW should be destined for deep disposal. Three options are being considered; a deep repository on land, a similar facility offshore with access by tunnel from the coast, and a deep offshore repository on the UK continental shelf, comprising either a mined facility or a group of deep boreholes, with access from a floating platform or artificial island. At the time of writing all of these options were being assessed at the conceptual level only, although a group of geological environments considered suitable for a deep onshore repository had been identified, and the areas which are potentially most suitable for disposal had been defined (Chapman et al, 1986b).

The impetus which led to the geological research programme in the UK, particularly into the feasibility of deep disposal of HLW, came from the 1976 report of the Royal Commission on Environmental Pollution (Cmnd 6618). The large-scale research programme continued until late in 1981 when the government decided to abandon all field studies into HLW disposal in favour of long-term storage of vitrified waste, on the basis that the principle of deep geological disposal had been demonstrated. This controversial decision led to the formulation of the current position in the 1982 White Paper, outlined above. A general review of the results of the curtailed geological research programme between 1976 and 1981 is provided by Mather *el al* (1982).

In March 1986, the House of Commons Environment Committee issued a report (HMSO, 1986) which was highly critical of the level of research into and technological implementation of waste disposal in the UK. It recommended, among other things, a review of the requirement for reprocessing spent fuel, more research into deep disposal of HLW/spent fuel, the development of an underground research laboratory at a 'non-repository' site, and various organizational changes involving the responsibilites of DoE, NIREX and RAWMAC. None of these suggestions was taken up by the government in its response to the report, although some minor policy changes resulted. (Cmnd 9852, 1986).

APPENDIX III

The Radioactive Waste Literature

Unfortunately for the general reader, much of the most useful and up-to-date basic reference material on radioactive waste disposal is contained in what is known as the 'grey literature': reports and papers issued by various national laboratories, with only limited circulation. Although these are generally freely available, through university or major national lending libraries, the problem is more one of finding out what has been written, since the reports are usually unadvertised. The reader will find many references to such reports in this book; laboratories and organizations such as Oak Ridge National Laboratory, Sandia, Lawrence Livermore, Battelle, Los Alamos and the Nuclear Regulatory Commission in the United States; the Swedish Nuclear Fuel Supply Company (SKB), their Swiss equivalent, NAGRA, Risö in Denmark, and Harwell and the British Geological Survey in the UK, all produce information in this form.

One of the easiest means of scanning through what is being released is to use one of the major abstracting publications, such as INIS (International Nuclear Information System) which is operated by the IAEA, and which regularly compiles almost all the published literature, including such grey material. In the UK, the Harwell Information Service prints a twice-monthly 'Information Bulletin on Radioactive Wastes and Fuel Reprocessing', which is taken from the INIS Atomindex data files. This abstracts about 300 reports and publications each month, and is the most convenient UK reference source.

As far as published and easily available material is concerned, there are two international journals devoted to the topic; *Radioactive Waste Management and the Nuclear Fuel Cycle*, and *Nuclear and Chemical Waste Management*, in addition to the many better known journals covering the fields of nuclear energy, chemistry, the earth sciences and health physics.

For a broad overview of the field it is probably best to go directly to the many booklets and conference proceedings edited or prepared by organizations such as the IAEA, the ICRP, the OECD Nuclear Energy Agency and the Commission of the European Communities.

241

The IAEA (Vienna) publishes 'guidebooks' in its Safety Series, and its Technical Report Series, which are written by groups of consultants and cover almost all aspects of the subject in a clear and general fashion. The ICRP publishes detailed and considered Recommendations on various aspects of radiological protection. At present the CEC is publishing the detailed results of most of the 1980–84 programme projects in the EUR Report Series (Luxembourg). All the organizations mentioned above sponsor large international conferences, and publish the proceedings in book form. The OECD-NEA (Paris) have convened many expert meetings over the past 15 years and publish a series of books, often in collaboration with other national or international bodies, on technical aspects which are of interest to member nations. For the yet more technically-minded, the US Materials Research Society sponsors an annual conference on 'The Scientific Basis for Nuclear Waste Management', which covers many of the detailed processes discussed in Chapters 5–7 of this book. To date nine volumes of proceedings of this conference have been published.

There are several other major international meetings held on a regular basis, including the annual 'Waste Management' series, held in Tucson, Arizona, which covers in considerable detail all aspects of work in the USA, as well as including invited up-date papers on the status of the main non-US programmes. This meeting is sponsored by a number of American organizations, including the American Nuclear Society and the American Society of Mechanical Engineers. In the UK a similar meeting is now sponsored by the British Nuclear Energy Society.

REFERENCES

Abelin, H., Moreno, L., Tunbrant, J. and Neretnieks, I. (1985). Flow tracer movement in some natural fractures in the Stripa granite—results from the Stripa Project, Phase 1, in *In-Situ Experiments in Granite Associated with the Disposal of Radioactive Waste*, OECD/NEA, Paris, pp. 67–81.

Abelin, H., Birgerson, L., Gidlund, J., Moreno, L., Neretneiks, I. and Tunbrant, S. (1986). Flow and tracer experiments in crystalline rocks. Results from several Swedish *in situ* experiments, in Werme, L. (ed.), *Scientific Basis for Nuclear Waste Management*, vol. 9, Materials Research Society, pp 627–639.

Allard, B. (1983). Actinide solution equilibria and solubilities in geological systems, KBS-83-35, SKBF/KBS, Stockholm.

Amarantos, S., de Batist, R., Brodersen, K., Glasser, F. P., Pottier, P. E., Vejmelka, R. and Zamorani, E. (1985). Behaviour of intermediate level waste forms in aqueous environments, in Simon, R. (ed.), *Radioactive Waste Management and Disposal*, Cambridge University Press, pp. 252–274.

Anderson, D M. (ed.), (1984). Smectite alteration. KBS-84-11, SKBF/KBS, Stockholm.

Apted, M. J. and Myers, J. (1982). Comparison of the hydrothermal stability of simulated spent fuel and borosilicate glass in a basaltic environment, Rockwell Hanford Operations Rpt. No. RHO-BW-ST-38P, 101 pp.

Apted, M J. Alexander, D. H., Liebtrau, A. M., Van Luik, A. E., Williford, R. E., Doctor, P. G. (1985). A conceptual model for repository source term evaluation, 82. Storage of high-level radioactive waste, *Nucl. Energy*, **21**, 245–252.

Atkinson, A. (1985). The time-dependence of the pH within a repository for radioactive waste disposal, AERE R.11777, HMSO, London.

Atkinson, A., Goult, D. J. and Hearn, J. A. (1986). An assessment of the long-term durability of concrete in radioactive waste repositories, in Werme, L. (ed.), *Scientific Basis for Nuclear Waste Management*, vol. 9, Materials Research Society, pp 239–246.

Barker, J. A. (1982). Laplace transform solutions for solute transport in fissured aquifers, *Adv. Water Resources*, **5**, 98–104.

Beale, H. (1982). Storage of high-level radioactive waste, *Nucl. Energy*, **21**, 245–252.

Beale, H., Engelman, H. J., Souquet, G., Mayence, M. and Hamstra, J. (1980). Conceptual design of repository facilities, in Simon and Orlowski (1980), pp. 488–512.

Beall, G. W. and Allard, B. (1982). Chemical aspects governing the choice of backfill materials for nuclear waste repositories, *Nucl. Technol.*, **59**, 405–408.

Bear, J. (1972). *Dynamics of Flow in Porous Media*, Elsevier, New York, 764 pp.

Bechai, M. and Heystee, R. J. (1986). Radioactive waste management in shallow tunnels in glacial till or clayey soil: geotechnical and hydrogeological considerations, in *Proc. International Symposium on the Siting Design and Construction of Underground Repositories for Radioactive Wastes*, IAEA, in press.

243

244

Berman, L. E., Ensminger, D. A., Gioffre, M. S. Koplik, C. M., Oston, S. G., Pollak, G. D. and Ross, B. I. (1978). Analysis of some nuclear waste management options, UCRL-13917, University of California, Berkeley.

Birgersson, L. and Neretnieks, I. (1984). Diffusion in the matrix of granitic rock: field test in the Stripa mine; part 2, in McVay, G. L. (ed.), *Scientific Basis for Nuclear Waste Management*, vol. 7, North Holland, pp. 247–254.

Birgersson, L., Abelin, H., Gidlund, J. and Neretnieks, I. (1985). Water flow rates and tracer transport in the 3-D drift in Stripa, in NEA (1985a), pp. 82–94.

Blacic, J. D. (1981). Importance of creep failure of hard rock in the near-field of a nuclear waste repository, in *Near-field Phenomena in Geologic Repositories for Radioactive Wastes*, OECD/NEA, Paris, pp. 121–129.

Black, J. H. and Barker, J. (1981a). Hydrogeological reconnaissance study: Worcester basin. Rpt. Inst. Geol. Sci., ENPU 81-3, British Geological Survey, Nottingham.

Black, J. H. and Barker, J. (1981b). Transient hydraulic tests in granite: fissured porous medium analysis and results, in Topp, S. V. (ed.), *Scientific Basis for Nuclear Waste Management*, vol. 6, North Holland, pp. 223–230.

Black, J. H. and Chapman, N. A. (1981). In search of nuclear burial grounds, *New Scientist*, **91**, 402–404.

Black, J. H., Holmes, D. C., Alexander, J. and Brightman, M. A. (1985). The role of low permeability rocks in regional flow systems: the Harwell study area, in IAH 17th Internat. Congress, '*Hydrogeology of rocks of low permeability*', Tucson, Arizona, pp. 107–117.

Boliden WP—Contech AB (1985). NAK WP-Cave Project. Report on the research and development stage May 1984 to October 1985, SKN Report 16, Stockholm.

Boult, K. A., Dalton, J. T., Hall, A. R., Hough, A. and Marples, J. A. C. (1978). The leaching of radioactive waste storage glasses, AERE-R9188, HMSO, London.

Bourke, P. J. and Hodgkinson, D. P. (1977). Granitic depository for radioactive waste— size, shape and depth v. temperatures. AERE-M-2900, HMSO, London, 21 pp.

Bourke, P. J. and Robinson, P. C. (1981). Comparison of thermally induced and naturally occurring water-borne leakages from hard-rock depositories for radioactive waste, *Rad. Waste Management*, **1**, 365–380.

Bradbury, M. H., Lever, D. and Kinsey, D. (1982). Aqueous phase diffusion in crystalline rock, in Lutze, W. (ed.), *Scientific Basis for Nuclear Waste Management*, vol. 5, North Holland, pp. 569–578.

Bradbury, M. H. and Stephen, I. G. (1986). Diffusion and permeability based sorption measurements in intact rock samples, in Werme, L. (ed.), *Scientific Basis for Nuclear Waste Management*, vol. 9, Materials Research Society, pp. 81–90.

Brereton, N. R. and Hall, D. H. (1983). Groundwater discharge mapping by thermal infra-red linescan surveying, Rpt. Inst. Geol. Sci., FLPU 83-7, British Geological Survey, Nottingham.

Brightman, M. A., Bath, A. H., Cave, M. R. and Darling, W. G. (1985). Pore fluids from the argillaceous rocks of the Harwell region, Rpt. Brit. Geol.Surv., FLPU 85-6, British Geological Survey, Nottingham.

Brookins, D. G. (1976). Shale as a repository for radioactive waste: evidence from Oklo, *Environ. Geol.*, **1**, 255–259.

Brookins, D. G. (1984). *Geochemical Aspects of Radioactive Waste Disposal*, Springer-Verlag, New York, 347 pp.

Burkholder, H. C., Cloninger, M. O., Baker, D. A. and Jansen, G. (1976). Incentives for partitioning high-level waste, *Nucl. Technol.*, **31**, 202–217.

Burton, W. R. and Griffin, J. R. (1980). A design study of long-term storage and underground disposal systems for highly active waste, Rpt. United Kingdom Atomic Energy Authority ND-R514 (R), UKAEA, Risley.

Cadelli, N., Cottone, G., Bertozzi, G. and Girardi, F. (1984). PAGIS: Summary report of Phase 1, Commission of the European Communities, EUR 9220 EN, Luxembourg.

Cartwright, K., Miller, J. R. and Berg, R. C. (1986). Hydrogeological experience at a low-level waste—shallow land burial site: a look toward the future, in *Geotechnical and Geohydrological Aspects of Waste Management*, A. A.Balkema, Rotterdam, pp. 63–79.

Castaing, R. (1983). Second Report of the Castaing Commission; MR1/CSSN. Rapport du groupe de travail sur les recherches et développement en matiere de gestion des déchets radioactifs proposé par le CEA,.Paris.

CEC (1982). Admissible Thermal Loadings in Geological Formations. Consequences on Radioactive Waste Disposal Methods, Commission of the European Communities, EUR 8179 EN/FR, Luxembourg.

Chan, T. and Cook, N. G. W. (1979). Calculated thermally induced displacements and stresses for heater experiments at Stripa, Sweden, LBL-7061, Lawrence Berkeley Laboratory.

Chapman, N. A. (1980). Mineralogical and geochemical constraints on maximum admissible repository temperatures, in *Underground disposal of radioactive wastes*, vol. 2, IAEA, Vienna, pp. 209–222.

Chapman, N. A., McKinley, I. G., Savage, D. and West, J. M. (1982). Mechanisms of dissolution of radioactive waste storage glasses and caesium migration from a granite repository, in Topp, S. V. (ed.), *Scientific Basis for Nuclear Waste Management*, vol. 6, North Holland, pp. 347–354.

Chapman, N. A., McKinley, I. G. and Smellie, J. A. T. (1984). The potential of natural analogues in assessing systems for deep disposal of high-level radioactive waste. NAGRA NTB-84-41, Baden, Switzerland, EIR BER NR 545, Würenlingen, and KBS TR-84-16, Stockholm, 103 pp.

Chapman, N. A. and Gera, F. (1985). Disposal of radioactive wastes in Italian clays: mined repository or deep boreholes? *Rad. Waste Manage. Nucl. Fuel Cycle*, **6**, 51–78.

Chapman, N. A. and Flowers, R. H. (1986). Near-field solubility constraints on radionuclide mobilization and their influence on waste package design, *Phil. Trans. R. Soc., Lond. A*, **319**, 83–95.

Chapman, N. A., Gera, F., Mittempergher, M. and Tassoni, E. (1986a). Disposal of radioactive wastes in Italian argillaceous formations, in *Siting, Design and Construction of Underground Repositories for Radioactive Wastes*, IAEA, Vienna, in press.

Chapman, N. A., McEwen, T. J. and Beale, H. (1986b). Geological environments for deep disposal of intermediate level waste in the United Kingdom, in *Siting, Design and Construction of Underground Repositories for Radioactive Wastes*, IAEA, Vienna, in press.

Choppin, G. R. and Rydberg, J. (1980). *Nuclear Chemistry*. Pergamon, Oxford.

Cmnd 6618 (1976). Nuclear power and the environment, Royal Commission on Environmental Pollution, Sixth Report, HMSO, London.

Cmnd 6820 (1977). Nuclear power and the environment, the government's response to the Sixth Report of the Royal Commission on Environmental Pollution, HMSO, London.

Cmnd 8607 (1982). Radioactive waste management, HMSO, London.

Cmnd 9852 (1986). *Radioactive Waste. The Government's Response to the Environment Committee's Report*. HMSO, London, 28 pp.

Côme, B. (1985). MIRAGE Project: Second Summary Progress Report (January to December 1984), EUR 10023 EN, Commission of the European Communities, Luxembourg, 177 pp.

Côme, B., Johnston, P. and Muller, A. (1985). *Design and Instrumentation of* in situ *Experiments in Underground Laboratories for Radioactive Waste Disposal*, A. A. Balkema, Rotterdam, 474 pp.

Côme, B. and Chapman, N. A. (1986). Natural Analogue Working Group: Report of the First Meeting, Bruxelles. EUR 10315 EN, Commission of the European Communities, Luxembourg.

Coons, W., Moore, E. L., Smith, M. J. and Kaser, J. D. (1980). The functions of an engineered barrier system for a nuclear waste repository in basalt, RHO-BWI-LD-23, Rockwell Hanford Operations, Richland, Wa.

246

Courbouleix, S. *et al.* (1985). *Etude géoprospective d'un site de stockage*, 5 vols, EUR 9866 FR, Commission of the European Communities, Luxembourg.

Curtis, D. B. and Gancarz, A. J. (1983). Radiolysis in nature: evidence from the Oklo natural reactors. KBS TR 83-10, Swedish Nuclear Fuel Supply Co.(SKB), Stockholm.

Deane, J. S. and Hollis, A. A. (1979). Practical aspects of heat transfer in radioactive waste repository design, AERE-R9343, HMSO, London.

Delcoigne, G. (1985). Trends in the nuclear controversy, 1980–84, *Nucl. Spectrum*, **1**, 15–19.

DoE (1984). Disposal facilities on land for low and intermediate level radioactive wastes: Principles for the protection of the human environment, HMSO, London.

DoE (1986). Assessment of Best Practicable Environmental Options (BPEOs) for management of low and intermediate level solid radioactive wastes, HMSO, London.

Dormuth, K. W. and Sherman, G. R. (1981). SYVAC—A computer program for assessment of nuclear fuel waste management systems, incorporating parameter variability, AECL-6814, Atomic Energy of Canada Ltd, Chalk River.

Duguid, J. O. (1979). Hydrogeologic transport of radionuclides from low-level waste burial grounds, in *Management of low-level radioactive waste*, vol. 2, Pergamon Press, Oxford, pp. 1119–1138.

Duncan, A. G. and Brown, S. R. A. (1982). Quantities of waste and a strategy for treatment and disposal, *Nucl. Energy*, **21**, 161–166.

Eiesenbud, M., Krauskopf, K., Penna Franca, E., Lei, W., Ballad, R., Linsalata, P. and Fujimori, K. (1984). Natural analogues for the transuranic actinide elements: an investigation in Minas Gerais, Brazil, *Environ. Geol. Water Sci.*, **6**, 1–9.

ELSAM/ELKRAFT (1981). *Disposal of high-level waste from nuclear power plants in Denmark*, 5 vols, ELSAM, Frederica, Denmark.

EPRI (1979). Status report on risk assessment for nuclear waste disposal, NP-1197, Palo Alto, California.

Faussat, A. (1985). The long-term management policy for low and intermediate level waste in France, in *Radioactive Waste Management*, British Nuclear Energy Society, London, pp. 99–102.

Forsyth, R. S., Werme, L. O. and Bruno, J. (1986). The corrosion of spent UO_2 fuel synthetic groundwater, *J. Nucl. Materials*, in press.

Fox, R. W. (1977). Ranger Uranium Environmental Inquiry, Second Report, Australian Government Publishing Service, Canberra.

Freeze, R. A. and Cherry, J. A. (1979). *Groundwater*, Prentice-Hall, New Jersey, 604 pp.

Friedlander, G., Kennedy, J. W. E., Macias, E. S. and Miller, J. M. (1981). *Nuclear and Radiochemistry*, 3rd edn, Wiley Interscience, New York.

Fritz, P. and Fontes, J. Ch. (1980). *Handbook of Environmental Isotope Geochemistry*, Vol. 1, Elsevier, Amsterdam.

Gera, F. (1972). Review of salt tectonics in relation to the disposal of radioactive wastes in salt formations, *Bull. Geol. Soc. Am.*, **83**, 3551–3574.

Goodwin, B. W. (1985). *The SYVAC approach for long-term environmental assessments*. Atomic Energy of Canada Ltd., Pinawa, Manitoba.

Gray, D. A., Greenwood, P. B., Bisson, G., Cratchley, C. R., Harrison, R. K., Mather, J. D., Poole, E. G., Taylor, B. J. and Willmore, P. L. (1976). Disposal of highly-active, solid radioactive wastes into geological formations—relevant geological criteria for the United Kingdom, Rpt. Inst. Geol. Sci. 76/12, HMSO, London.

Gray, W. J. and McVay, G. L. (1983). Comparison of spent fuel and UO_2 release in salt brines, in Werme, L. (ed.), *Proc. Third Spent Fuel Workshop*, KBS 83–76, Stockholm.

Greenhalgh, J. R., Fell, T. P. and Adams, N. (1985). Doses from intakes of radionuclides by adults and young people, National Radiological Protection Board, NRPB-R162, HMSO, London.

Grisak, G. E. and Pickens, J. F. (1980). Solute transport through fractured media, I: the effect of matrix diffusion, *Water Resour. Res.*, **16**, 719–730.

Grogan, H. A. (1985). Concentration ratios for BIOPATH—selection of the soil-to-plant database, NAGRA NTB-85-16, NAGRA, Baden, Switzerland.

Hatch, L. P. (1953). Ultimate disposal of radioactive wastes, *Am. Sci.*, **41**, 410–421.

Henrion, P. N., Monsecour, M., Fonteyne, A., Put, M. and de Regge, P. (1985). Migration of radionuclides in Boom Clay, *Rad. Waste Manag. Nuclear Fuel Cycle*, **6**, 313–359.

Herbert, A. W. (1985). The verification of NAMMU using HYDROCOIN Level 1 Cases 1 and 7: transient flow from a borehole and saturated flow through a shallow land disposal facility. AERE-R-11944, HMSO, London, 17 pp.

Herbert, A. W., Hodgkinson, D. P., Lever, D. A., Rae, J. and Robinson, P. C. (1984). Mathematical modelling of radionuclide migration in groundwater. AERE-TP-1087. United Kingdom Atomic Energy Authority, Harwell, 21 pp.

Hill, M. D. (1979). Analysis of the effects of variations in parameter values on the predicted radiological consequences of geological disposal of high-level waste, NRPB-R86, National Radiological Protection Board, HMSO, London.

Hill, M. D. and Grimwood, P. D. (1978). Preliminary assessment of the radiological protection aspects of disposal of high-level waste in geological formations, NRPB-R69, National Radiological Protection Board, HMSO, London.

Hill, M. D. and Lawson, G. (1980). An assessment of the radiological consequences of disposal of high-level waste in coastal geological formations, NRPB-R108, National Radiological Protection Board, HMSO, London.

Hill, M. D. *et al.* (1981). An assessment of the radiological consequences of disposal of intermediate-level wastes in argillaceous rock formations, NRPB-R126, National Radiological Protection Board, HMSO, London.

Hill, M. D. and Webb, G. A. M. (1985). Radiological protection criteria for waste management, *Nucl. Energy*, **24**, 111–115.

Hill, M. D. and Smith, G. M. (1986). Does optimisation have a role in radioactive waste management decisions? in *Optimisation of Radiological Protection*, IAEA, Vienna, in press.

HMSO, (1986). Radioactive Waste. First Report of the Environment Committee (Chairman, Sir Hugh Rossi), 1985–6 session, vol. 1: report and proceedings of the committee, House of Commons Paper 191–1, 174 pp.

Hodgkinson, D. P. (1977). Deep rock disposal of high-level radioactive waste: transient heat conduction from dispersed blocks, AERE-M2900, HMSO, London.

Hodgkinson, D. P. (1980). A mathematical model for hydrothermal convection around a radioactive waste depository in hard rock, *Ann. Nucl. Energy*, **7**, 313–334.

Hodgkinson, D. P. and Bourke, P. J (1980). Initial assessment of the thermal stresses around a radioactive waste depository in hard rock, *Ann. Nucl. Energy*, **7**, 541–552.

Holliday, F. G. T. (1984). The report of the independent review of disposal of radioactive waste in the North-East Atlantic, HMSO, London.

Holmes, D. C. (1981). Hydraulic testing of deep boreholes at Altnabreac: development of the testing system and initial results, Rpt. Inst. Geol. Sci., ENPU 81–4, British Geological Survey, Nottingham.

Hood, M. (1979). Some results from a field investigation of thermo-mechanical loading of a rock mass when heated canisters are emplaced in the rock, LBL-9392, Lawrence Berkeley Laboratories.

IAEA (1977). Site selection factors for repositories of solid high-level and alpha-bearing wastes in geological formations, Technical Report Series No 177, International Atomic Energy Agency, Vienna, 64 pp.

IAEA (1978). Proceedings of the technical committee meeting on natural fission reactors, STI/PUB/475, International Atomic Energy Agency, Vienna.

IAEA (1981). Shallow ground disposal of radioactive wastes: a guidebook, Safety Series No. 53, International Atomic Energy Agency, Vienna.

IAEA (1982a). Site investigations for repositories for solid radioactive wastes in deep continental geological formations, Tech. Rpt. Series, No. 215, International Atomic Energy Agency, Vienna, 106 pp.

IAEA (1982b). Site investigations for repositories for solid radioactive wastes in shallow ground, Tech. Rpt. Series No. 216, International Atomic Energy Agency, Vienna, 89 pp.

IAEA (1983a). Concepts and examples of safety analyses for radioactive waste repositories in continental geological formations, Safety Series No. 58, International Atomic Energy Agency, Vienna, 171 pp.

IAEA (1983b). Disposal of radioactive grouts into hydraulically fractured shale, Tech. Rpt. Series No. 232, International Atomic Energy Agency, Vienna, 111 pp.

IAEA (1985). Techniques for site investigations for underground disposal of radioactive wastes Tech. Rpt. Series No. 256, International Atomic Energy Agency, Vienna, 62 pp.

ICRP (1977). Recommendations of the International Commission on Radiological Protection, ICRP Publn, No. 26, Ann. ICRP, vol. 1 (3), Pergamon, Oxford.

ICRP (1985). Radiation protection principles for the disposal of solid radioactive waste, ICRP Publn, No. 46, Ann. ICRP, vol. 15 (4), Pergamon, Oxford.

Ivanovich, M. and Harmon, R. S. (1982). Uranium series disequilibrium, Clarendon, Oxford, 571 pp.

Jeffrey, J. A., Chan, T., Cook, N. G. W. and Witherspoon, P. A. (1979). Determination of *in situ* thermal properties of Stripa granite from temperature measurements in the full-scale heater experiments, LBL-8423, Lawrence Berkeley Laboratories.

Jenne, E. A. (1981). Geochemical modelling: a review, PNL-3574, Battelle Pacific Northwestern Laboratories, Richland.

Jensen, B. S. (1982). *Migration Phenomena of Radionuclides into the Geosphere.* Harwood Academic, 197 pp.

Johnson, L. H. and Wikjord, A. G. (1981). The rate of mobilisation of radionuclides from nuclear fuel and reprocessed wastes, AECL-TR-79, 91–104, Atomic Energy of Canada Ltd, Pinawa, Manitoba.

Kane, P. and Thorne, M. C. (1984). User's guide to the biosphere code ECOS, DoE SYVAC TN-ANS-9, Department of the Environment, London.

KBS (1977). *Handling of Spent Nuclear Fuel and Final Storage of Vitrified High-level Reprocessing Wastes,* 5 vols, Swedish Nuclear Fuel Supply Co. (SKBF/KBS), Stockholm.

KBS (1978). *Handling and Final Storage of Unreprocessed Spent Nuclear Fuel,* 2 Vols, Swedish Nuclear Fuel Supply Co. (SKBF/KBS), Stockholm.

KBS (1983). *Final Storage of Spent Nuclear Fuel—KBS3,* 4 vols, Swedish Nuclear Fuel Supply Co. (SKBF/KBS), Stockholm.

Kelleher, W. J. (1979). Water problems at the West Valley burial site, in *Management of Low-level Radioactive Waste,* vol. 2, Pergamon, Oxford, pp. 843–851.

Kelsall, P. C., Case, J. B. and Chabannes, C. R. (1984). Evaluation of excavation induced changes in rock permeability, *Int. J. Rock Mech. Min. Sci. & Geochem. Abstr.,* **21,** 123–135.

Lake, L. M. *et al.* (1985). Backfilling and sealing of repositories and access shafts and galleries in clay, granite and salt formations, in Simon, R. (ed.), *Radioactive Waste Management and Disposal,* Cambridge University Press, 562–574.

Larsson, A. H., Andersson, K. A., Grundfelt, B. and Hadermann, J. (1983). Mathematical models for nuclide transport in geological media—an international comparison (INTRACOIN), in *Radioactive Waste Management',* Proceedings of Intl. Conf., Seattle. IAEA, Vienna, vol. 5, pp. 197–212.

Lasch, M., Schaller, K. H., Stang, W. and Watzel, G. V. P. (1984). *Decommissioning of Nuclear Power Plants,* Proceedings of European Conf. Rad. Waste Management, Luxembourg, Graham and Trotman, London.

Lawson, G. and Smith, G. M. (1985). BIOS: a model to predict radionuclide transfer and doses to man following releases from geological repositories for radioactive waste, NRPB-R169, HMSO, London.

Lindberg, R. D. and Runnels, D. D. (1984). Groundwater redox reactions: an analysis of equilibrium state applied to Eh measurements and geochemical modelling, *Science*, **225**, 925–7.

Marples, J. A. C., Lutze, W. and Sombret, C. (1980). The leaching of solidified high-level waste under various conditions, in: Simon and Orlowski (1980), 307–323.

Marsh, G. P. (1982). Materials for high-level waste containment, *Nucl. Energy*, **21**, 253–266.

Marsily, G. de (1986). *Quantitative Hydrogeology* (*Groundwater Hydrology for Engineers*). Academic, New York, 450 pp.

Marsily, G. de and Merriam, D. F. (1982). *Predictive Geology*, Pergamon, Oxford, 206 pp.

Mather, J. D., Chapman, N. A., Black, J. H and Lintern, B. C. (1982). The geological disposal of high-level radioactive waste—a review of the Institute of Geological Sciences' research programme, *Nucl. Energy, 21*, 167–174.

Matthews, M. L. (1986). UMTRA Project: Implementation of design, in *Geotechnical and Geohydrological Aspects of Waste Management*, A. A. Balkema, Rotterdam, pp. 15–21.

McCarthy, G. J. (1976). High-level waste ceramics, *Trans. Am. Nucl. Soc.*, **23**, 168–169.

McKinley, I. G. (1985a). The quantification of source-term profiles from near-field geochemical models, in Proceedings of Workshop on the source term for radionuclide migration from high-level waste or spent fuel under realistic repository conditions, Sandia Rpt., SAND 85-0380, pp. 115–123.

McKinley, I. G. (1985b). The geochemistry of the near-field, NAGRA NTB-84-48, NAGRA, Baden, Switzerland.

McKinley, I. G. and Hadermann, J. (1984). Radionuclide sorption database for Swiss safety assessments, NAGRA NTB 84-40, Baden, Switzerland.

McKinley, I. G., West, J. M. and Grogan, H. A. (1985). An analytical overview of the consequences of microbial activity in a Swiss HLW repository, NAGRA NTB 85–43, Baden, Switzerland.

Milodowski, A. E., George, I. A., Bloodworth, A. J. and Robins, N. S. (1985). Reactivity of ordinary Portland cement (OPC) grout and various lithologies from the Harwell research site, Rpt. Brit. Geol. Survey, FLPU 85-15, 34 pp.

Montgomery, D. M. and Blanchard, R. L. (1979). Radioactivity measurements in the environment of the Maxey Flats waste burial site, in *Management of Low-level Radioactive Waste*, vol. 2, Pergamon, Oxford, pp. 763–786.

Morgan, P. E. D., Clarke, D. R., Jantzen, C. M. and Harker, A. B. (1981). High-alumina tailored nuclear waste ceramics, *Jour. Am. Ceram. Soc.*, **64**, 249–258.

Morrow, G., Lockner, D., Moore, D. and Byerlee, J. (1981). Permeability of granite in a temperature gradient, *J. Geophys. Res.*, **86**, 3002–3008.

NAGRA (1983). Bitumen, ein Verfestigungsmaterial fur radioaktive Abfalle und seine historischen Analoga, NAGRA NTB 83-11, Baden, Switzerland.

NAGRA (1985). *Project Gewähr 1985*, 8 vols in German. Summary volume in English, NAGRA NGB-85-09, NAGRA, Baden, Switzerland.

NAS/NRC (1978). Geological criteria for repositories for high-level radioactive wastes, Natl. Acad. Sci., Washington DC.

NEA (1980). *Borehole and Shaft Plugging*, Proceedings of a workshop, Columbus, Ohio, OECD Nuclear Energy Agency, Paris, 434 pps.

NEA (1984a). *Geological Disposal of Radioactive Waste. An Overview of the Current Status of Understanding and Development*, OECD Nuclear Energy Agency, Paris, 116 pp.

NEA (1984b). *Long term Radiation Protection Objectives in Radioactive Waste Disposal*, OECD Nuclear Energy Agency, Paris.

NEA (1985a). In situ *Experiments in Granite Associated with the Disposal of Radioactive Waste*, OECD Nuclear Energy Agency, Paris, 266 pp.

NEA (1985b). Summary of nuclear power and fuel cycle data in OECD member countries, OECD Nuclear Energy Agency, Paris.

NEA (1985c). Review of the continued suitability of the dumping site for radioactive waste in the North-East Atlantic, OECD Nuclear Energy Agency, Paris.

Neretnieks, I. (1980). Diffusion in the rock matrix: an important factor in radionuclide retardation? *J. Geophys. Res.*, **85**, 4379–4397.

Neretnieks, I. (1982). Diffusivities of some dissolved constituents in compacted wet bentonite clay (MX80), and the impact on radionuclide migration in the buffer, KBS-82-27, Swedish Nuclear Fuel Supply Co., Stockholm.

Neretnieks, I. (1985). Some aspects of the use of iron canisters in deep lying repositories for nuclear waste, NAGRA NTB-85-35, NAGRA, Baden, Switzerland.

Noy, D. J. and Holmes, D. C. (1986). A single hole tracer test to determine longitudinal dispersion, Rpt. Brit. Geol.Surv., FLPU 86–5, British Geological Survey, Nottingham.

Ogard, A. E. and Bryant, E. A. (1982). The misused and misleading IAEA leach test, *Nucl. Chem. Waste Management*, **3**, 79–81.

ONWI, (1983). Very deep hole systems engineering studies, US Office of Nuclear Waste Isolation, Rpt. ONWI-226, Battelle Laboratories, Columbus, Ohio.

Oversby, V. M. and Ringwood, A. E. (1982). Leaching studies on SYNROC at 95°C and 200°C, *Rad. Waste Management*, **2**, 223–37.

Oversby, V. M. and McCright, R. D. (1985). Laboratory experiments designed to provide limits on the radionuclide source term for the NNWSI project, in Proceedings of the Workshop on the source term for radionuclide migration from high-level waste or spent fuel under realistic repository conditions, Sandia Rpt., SAND 85-0380, pp. 175–186.

Papadopolous, S. S. and Winograd, I. J. (1974). Storage of low-level radioactive wastes in the ground: hydrogeologic and hydrochemical factors, US Geol. Surv., Open-file Rpt. 74-344, 49 pp.

Parker, F. L., Broshears, R. E. and Pasztor, J. (1984). *The Disposal of High-level Radioactive Waste 1984*, 2 vols, Rpt NAK-11, National Board for Spent Nuclear Fuel, Stockholm.

Pinner, A. V., Hemming, C. R. and Hill, M. D. (1984). An assessment of the radiological protection aspects of shallow land burial of radioactive wastes, NRPB-R-161. National Radiological Protection Board, HMSO, London.

Pusch, R. (1979). Highly compacted sodium bentonite for isolating rock-deposited radioactive waste products, *Nuc. Tech.*, **45**, 153–157.

Pusch, R., Nilsson, J., and Ramqvist, G. (1985). Final Report of the Buffer Mass Test, Stripa Project Reports 85-11 and 85-12 (2 vols), SKB-Swedish Nuclear Fuel Supply Co., Stockholm.

Rae, J., Robinson, P. C. and Wickens, L. M. (1981). A user's guide for the program NAMMU, AERE-R10120, HMSO, London.

Rae, J., Robinson, P. C. and Wickens, L. M. (1983). Coupled heat and groundwater flow in porous rocks, in Lewis, R. W. *et al.* (eds), *Numerical Methods in Heat Transfer*, vol. 2, Wiley, Chichester.

Rasmuson, A. and Neretnieks, I. (1983). Surface migration in sorption processes, KBS-TR-83-37. Swedish Nuclear Fuel Supply Co., Stockholm.

RAWMAC, (1980) (First Report) onwards. Radioactive Waste Management Advisory Committee, Annual Reports, HMSO, London.

Relyea, J. F. and Serne, R. J. (1979). Controlled sample program publication number 2: interlaboratory comparison of batch Kd values, PNL-2872, Battelle Pacific Northwest Laboratories, Richland.

Ringwood, A. E., Kesson, S. E., Ware, N. G., Hibberson, W. O. and Major, A. (1979). The SYNROC process: a geochemical approach to nuclear waste immobilisation, *Geochem. J.*, **13**, 141–165.

Robertson, J. B. (1980). Shallow land burial of low-level radioactive wastes in the USA, in *Underground Disposal of Radioactive Wastes*, vol. 2, IAEA, Vienna, pp. 253–269.

Robins, N. S., Martin, B. A. and Brightman, M. A. (1981). Borehole drilling and completion: details for the Harwell research site. *Rpt. Brit. Geol. Surv.* ENPU 81–9, British Geological Survey, Nottingham, 28 pp.

Robinson, P. C. (1983). Connectivity of fracture systems—a percolation theory approach. *J. Phys. A: Math. Gen.*, **16**, 605–614.

Rogers, V. C. (1986). Low-level waste: The challenge of disposal, in *Geotechnical and Geohydrological Aspects of Waste Management*, A. A. Balkema, Rotterdam, pp. 51–61.

Saltelli, A., Bertozzi, G. and Stanners, D. A. (1984). LISA: a code for safety assessment in nuclear waste disposal, EUR-9306, Commission of the European Communities, Brussels.

Savage, D. (1984). The geochemical interactions of simulated borosilicate waste glass, granite and water at 100–350°C and 50MPa, Rpt. Brit. Geol. Survey, FLPU 84-3, British Geological Survey, Keyworth, 77 pp.

Savage, D. and Chapman, N. A. (1982). Hydrothermal behaviour of simulated waste glass, and waste-rock interactions under repository conditions, *Chem. Geol.*, **36**, 59–86.

Savage, D. and Robbins, J. E. (1983). The interaction of borosilicate glass and granodiorite at 100°C, 50MPa: implications for models of radionuclide release, in Lutze, W. (ed.), *Scientific basis for Nuclear Waste Management*, vol. 7, Elsevier, Amsterdam, pp. 145–152.

Schweingruber, M. (1983). Actinide solubility in deep groundwaters—estimates for upper solubility limits based on chemical equilibrium calculations, NAGRA NTB 83-24, Baden Switzerland.

Simon, R. (ed.) (1985). *Radioactive Waste Management and Disposal*, Proceedings of the Second European Community Conference, Luxembourg, Cambridge University Press, 734 pp.

Simon, R. and Orlowski, S. (eds) (1980). *Radioactive Waste Management and Disposal*, Proceedings of the First European Community Conference, Luxembourg, Harwood Academic Press, 693 pp.

Skagius, K. and Neretnieks, I. (1986). Diffusivities in crystalline rock materials, in Werme, L. (ed.), *Scientific Basis for Nuclear Waste Management*, vol. 9, Materials Research Society, pp. 73–80.

SKI, (1985). HYDROCOIN: Progress Report No. 2, January to June 1985, Swedish Nuclear Power Inspectorate (SKI), Stockholm, 17 pp.

Smyth, J. R. (1982). Zeolite stability constraints on radioactive waste isolation in zeolite-bearing volcanic rocks, *J. Geol.*, **90**, 195–201.

Stephansson, O., Blomquist, R., Groth, T., Jonasson, P. and Tarandi, T. (1979). Modelling of temperature fields and deformations for radioactive waste repositories in hard rock, in *Underground Disposal of Radioactive Wastes*, vol. 2, IAEA, Vienna, pp. 121–133.

Stephansson, O. and Groth, T. (1980). Modelling of rock mass deformation for radioactive waste repositories in hard rock, KBS-TR-80-02, Swedish Nuclear Fuel Supply Co., Stockholm.

Stevens, P. R. and Debuchananne, G. D. (1976). Problems in shallow land disposal of solid low-level radioactive waste in the United States, *Bull. Intern. Assn. Engin. Geol.*, **14**, 161–171.

Stroes-Gascoyne, Johnson, L. H., Beeley, P. A. and Sellinger, D. M. (1986). Dissolution of used CANDU fuel at various temperatures and redox conditions, in Werme, L. (ed.), *Scientific Basis for Nuclear Waste Management*, vol. 9, Materials Research Society, pp. 317–326.

Stumm, W. and Morgan, J. J. (1981). *Aquatic Chemistry*, 2nd edn, Wiley, New York.

Torstenfelt, B., Andersson, K., Allard, B. and Olofsson, U. (1982). Diffusion measurements in compacted bentonite, in Topp, S. V. (ed.), *Scientific Basis for Nuclear Waste Management*, vol. 4, North Holland, pp. 295–302.

UNEP (1985). Radiation, doses, effects, risks, United Nations Environment Programme, Nairobi.

UNSCEAR (1982). Ionising radiation: Forces and biological effects, Report of the United Nations Scientific Committee on the Effects of Atomic Radiation, UNSCEAR-8298, United Nations, New York.

Vandenbergh, N., Bonne, A., and Heremans, R. H. (1980). Scenarios d'évolution geologique lente, appliqués au site argileux de Mol (Belgique), in *Radionuclide Release Scenarios for Geologic Repositories*, OECD Nuclear Energy Agency, Paris, pp. 169–180.

Vandergraaf, T. T., Abry, D. R. M. and Davis, K. E. (1982). The use of autoradiography in determining the distribution of radionuclides sorbed in thin sections of plutonic rocks from the Canadian Shield, *Chem. Geol.*, **36**, 139–154.

Walton, R. D. and Cowan, G. A. (1975). Relevance of nuclide migration at Oklo to the problem of geological storage of radioactive waste, in *The Oklo Phenomenon*, International Atomic Energy Agency, Vienna, pp. 499–507.

Wang, R. and Katayama, Y. B. (1982). Dissolution mechanisms for UO_2 and spent fuel, *Nucl. Chem. Waste Management*, **3**, 83–90.

Webb, G. A. M. *et al.* (1986). Development of a general framework for the practical implementation of ALARA, in Proc. Int. Symp. *Optimisation of Radiological Protection*, IAEA, Vienna, in press.

Webster, D. A. (1979). Land burial of solid radioactive waste at Oak Ridge National Laboratory, Tennessee; a case history, in *Management of Low-level Radioactive Waste*, vol. 2, Pergamon, Oxford, pp. 731–746.

West, J. M., Christofi, N. and McKinley, I. G. (1985). An overview of recent microbial research relevant to the geological disposal of nuclear waste, *Rad. Waste Manage. Nucl. Fuel Cycle*, **6**, 79–95.

Williams, G. M. *et al.* (1985). *In situ* radionuclide migration studies in a shallow sand aquifer. Part I, FLPU 85-7; Part II, FLPU 85-10, Rpt. Brit. Geol. Surv., British Geological Survey, Nottingham.

Wood, M. I. and Coons, W. E. (1982). Basalt as a potential waste package backfill component in a repository located within the Columbia River basalt, *Nucl. Tecnol.*, **59**, 409–419.

Zussman, J. (ed.) (1977). *Physical Methods in Determinative Mineralogy*, 2nd edn, Academic, New York, 720 pp.

Index